SALTERS-NUFFIELD
AS/A level Biology

1

THIRD EDITION

ALWAYS LEARNING

PEARSON

Published by **Pearson Education Limited, 80 Strand, London WC2R 0RL**

www.pearsonschoolsandfecolleges.co.uk

Copies of official specifications for all Edexcel qualifications may be found on the website: www.edexcel.com

Text © University of York Science Education Group, 2015

Designed by Elizabeth Arnoux, Pearson Education Ltd
Typeset by Wearset Ltd, Boldon, Tyne and Wear
Original illustrations © Pearson Education
Illustrated by Pantek Arts, Maidstone, Kent; Peter Bull Art Studio; Wearset Ltd, Boldon, Tyne and Wear
Cover design by Juice Creative
Picture research by Jane Smith
Cover photo © Science Photo Library/Frans Lanting, Mint Images

The rights of Peter Anderson, Nick Owens, Cathy Rowell, David Slingsby, Mark Smith and Nicola Wilberforce to be identified as authors of this work have been asserted by them in accordance with the Copyright, Designs and Patents Act 1988.

First edition published 2005
Second edition published 2008
This edition published 2015
Published as a trial edition in 2002
19 18 17 16 15
10 9 8 7 6 5 4 3

British Library Cataloguing in Publication Data
A catalogue record for this book is available from the British Library

ISBN 978 1 447 99100 7

Printed in Slovakia by Neografia

Websites
Pearson Education Limited is not responsible for the content of any external internet sites. It is essential for tutors to preview each website before using it in class so as to ensure that the URL is still accurate, relevant and appropriate. We suggest that tutors bookmark useful websites and consider enabling students to access them through the school/college intranet.

From the Publisher

In order to ensure that this resource offers high-quality support for the associated Edexcel qualification, it has been through a review process by the awarding body to confirm that it fully covers the teaching and learning content of the specification or part of a specification at which it is aimed, and demonstrates an appropriate balance between the development of subject skills, knowledge and understanding, in addition to preparation for assessment.

While the publishers have made every attempt to ensure that advice on the qualification and its assessment is accurate, the official specification and associated assessment guidance materials are the only authoritative source of information and should always be referred to for definitive guidance.

Edexcel examiners have not contributed to any sections in this resource relevant to examination papers for which they have responsibility.

No material from an endorsed book will be used verbatim in any assessment set by Edexcel.

Endorsement of a book does not mean that the book is required to achieve this Edexcel qualification, nor does it mean that it is the only suitable material available to support the qualification, and any resource lists produced by the awarding body shall include this and other appropriate resources.

Many people from schools, colleges, universities, industries and the professions have contributed to the Salters-Nuffield Advanced Biology project. They include the following for this third edition.

Project editor

Anne Scott — University of York Science Education Group (UYSEG)

Authors

Peter Anderson	Ampleforth College
Nick Owens	
Cathy Rowell	Bootham School, York
David Slingsby	
Mark Smith	Grammar School at Leeds
Nicola Wilberforce	Esher College

Acknowledgements

We would also like to thank the following for their assistance in the development of these materials.

Teachers, technicians and students at schools and colleges running the Salters-Nuffield Advanced Biology course.

Joanna MacDonald (Project Administrator)   University of York Science Education Group

Sponsors

We are grateful for sponsorship from The Salters' Institute who have continued to support the Salters-Nuffield Advanced Biology project after its initial development and have enabled the production of these materials.

Authors of the previous editions

This revised edition of the Salters-Nuffield Advanced Biology course materials draws heavily on the initial project development and the work of previous authors.

Glen Balmer	Watford Grammar School	Paul Heppleston	
Susan Barker	Institute of Education, University of Warwick	Liz Jackson	King James's School, Knaresborough
Martin Bridgeman	Stratton Upper School, Biggleswade, Bedfordshire	Christine Knight	
Alan Clamp	Ealing Tutorial College	Pauline Lowrie	Sir John Deane's College, Northwich
Mark Colyer	Oxford College of Further Education	Peter Lillford	Department of Biology, University of York
Jon Duveen	City & Islington College, London	Jenny Owens	Rye St Antony School, Headington, Oxford
Brian Ford	The Sixth Form College, Colchester	Nick Owens	Oundle School, Peterborough
Richard Fosbery	The Skinners School, Tunbridge Wells	Michael Reiss	Institute of Education, University of London
Barbara Geatrell	The Burgate School, Fordingbridge, Hants	Cathy Rowell	Bootham School, York
Ginny Hales	Cambridge Regional College	Jamie Shackleton	Cambridge Regional College
Angela Hall	Nuffield Curriculum Centre	David Slingsby	Wakefield Girls High School
Steve Hall	King Edward VI School, Southampton	Mark Smith	Leeds Grammar School
Sue Howarth	Tettenhall College	Nicola Wilberforce	Esher College
Gill Hickman	Ringwood School	Jane Wilson	Coombe Dean School, Plymouth, Devon
Liz Hodgson	Greenhead College, Huddersfield	Mark Winterbottom	King Edward VI School, Bury St Edmunds
Laurie Haynes	School of Biological Sciences, University of Bristol		

Advisory Committee for the initial development

Professor R McNeill Alexander FRS	University of Leeds
Dr Roger Barker	University of Cambridge
Dr Allan Baxter	GlaxoSmithKline
Professor Sir Tom Blundell FRS (Chair)	University of Cambridge
Professor Kay Davies CBE FRS	University of Oxford
Professor Sir John Krebs FRS	Food Standards Agency
Professor John Lawton FRS	Natural Environment Research Council
Professor Peter Lillford CBE	University of York
Dr Roger Lock	University of Birmingham
Professor Angela McFarlane	University of Bristol
Dr Alan Munro	University of Cambridge
Professor Lord Robert Winston	Imperial College of Science, Technology and Medicine

Please cite this publication as: Salters-Nuffield AS/A level Biology Student book, Edexcel Pearson, London, 2015

A context-led approach

Salters-Nuffield Advanced Biology (SNAB) is a context-led course which embraces a student centred approach. Each topic uses a real-life context; a storyline or contemporary issue is presented, and as you work through the topic you learn relevant biological concepts to aid your understanding of the context. The opening pages of each topic introduce the context and provide an overview of the biological principles that are covered in the topic. For example, in Topic 1 you are introduced to Mark, a 15-year-old who had a stroke, and Peter, an adult who had a heart attack. You study the biological principles needed to understand what happened to Mark and Peter. SNAB combines the key concepts underpinning biology today, with the opportunity to gain the wider skills that biologists need in the twenty first century.

Building knowledge through the course

Salters-Nuffield Advanced Biology has been carefully designed so that you build up your understanding of biological ideas gradually. For example, there is not a section labelled 'biochemistry' containing everything you might need to know on carbohydrates, fats, nucleic acids and proteins. In SNAB you study the biochemistry of these large molecules bit by bit throughout the course when you need to know the relevant information for a particular topic. In this way information is presented in manageable chunks. As you progress through the course you revisit and extend ideas building on existing knowledge. For example in Topic 1 the structure and function of the heart and circulation are introduced, and in Topic 7, you learn more detail including the control of heart rate.

Activities as an integral part of the learning process

SNAB encourages an active approach to learning. Throughout this book you are directed to a wide variety of activities that can be found within the online resources. These have been designed to aid learning of both content and skills, including practical, mathematical and wider skills such as data analysis, critical evaluation of information, communication and collaborative work.

Within the electronic resources you will find a wealth of interactive materials including animations on such things as the cardiac cycle and cell division. These animations are designed to help you understand the more difficult bits of biology. There are also support sections that should be useful if you need help with biochemistry, mathematics, ICT, practical skills, study skills, and the examination.

Developing practical skills

Throughout the course there are opportunities for the development of practical skills. A structured approach to developing these skills has been carefully integrated into the resources, this includes a framework for you to use and reflect on the skills developed. Your knowledge and understanding of these skills will be assessed within the written assessment at the end of the course. Your competence in practical skills will be assessed as you progress through the course. See the online resources to find out more about the development of practical skills.

SNAB and ethical debate

With rapid developments in biological science, we are faced with an increasing number of challenging decisions. For example, the rapid advances in gene technology present ethical dilemmas. Should embryonic stems cells be used in medicine? Which genes can be tested for in prenatal screening?

In SNAB you develop the ability to discuss and debate these types of biological issues. There is rarely a right or wrong answer; rather you learn to justify your own decisions using ethical frameworks.

What you need to learn

The specification for Edexcel Biology A (Salters-Nuffield) defines what will be examined for SNAB. Each of the topics includes a check your notes activity. These allow you to make sure that you have grasped all the required knowledge and understanding for the topic.

Any questions?

If you have any questions or comments about the materials you can let us know via the website or write to us at:

The Salters-Nuffield Advanced Biology Project

University of York Science Education Group

Alcuin D

University of York

Heslington

York YO10 5DD

www.advancedbiology.org

There are features in the student books that will help your learning and help you find your way around the course.

This book covers the four topics that make up the AS course and the first year of the full A level course. These are shown in the contents list, which also shows you the page numbers for the main sections within each topic. There is an index at the back to help you find what you are looking for.

Main text

Key terms in the text are shown in **bold type**.

There is an introduction at the start of each topic. This introduces the context and provides a guide to the sort of things you will be studying in the topic.

There is an '**Overview**' box on the first spread of each topic, so you know which biological principles will be covered.

OVERVIEW

In this topic you will study how changes in DNA can result in genetic disease using the example of cystic fibrosis. You will first look in detail at the symptoms of cystic fibrosis, extending your previous knowledge . . .

Occasionally in the topics there are also '**Key biological principle**' boxes where a fundamental biological principle is highlighted.

KEY BIOLOGICAL PRINCIPLE

Living organisms have to exchange substances with their surroundings. For example, they take in oxygen and nutrients and get rid of waste materials such as carbon dioxide. In unicellular organisms the whole cell . . .

'**Did you know?**' boxes contain material that will not be examined, but we hope you will find it interesting.

DID YOU KNOW?

What are epithelial cells?

Epithelial cells form the outer surface of many animals including mammals. They also line the cavities and tubes within the animal and cover the surfaces of internal organs. The cells work together as a . . .

Questions

You will find two types of question in this book.

In-text questions occur now and again in the text. They are intended to help you to think carefully about what you have read and to aid your understanding. You can self-check using the answers provided at the back of the book.

CHECKPOINT

2.1 Describe the properties of gas exchange surfaces.

Boxes containing '**Checkpoint**' questions are found throughout the book. They give you summary-style tasks that build up some revision notes as you go through the student book. Your teacher will be able to provide you with the answers to these questions.

Links to the online resources

'**Activity**' boxes show you which activities are associated with particular sections of the book. Activity sheets and any related animation can be accessed from the activity homepages found via 'topic resources' in the SNAB Online resources. Activity sheets include such things as practicals, worksheets that accompany animations and interactive tutorials, issues for debate and role plays. They can be printed out. Your teacher or lecturer will guide you on which activity to do and when. There may also be weblinks associated with the activity, giving hotlinks to other useful websites.

A final activity for each topic enables you to '**check your notes**' using the topic summary provided within the activity. The topic summary shows you what you need to have learned.

'**Weblink**' boxes give you useful websites to go and look at.

'**Extension**' boxes refer you to extra information or activities available in the electronic resources. The extension sheets can be printed out. The material in them is not covered directly by the specification.

'**Support**' boxes are provided now and again, where it is particularly useful for you to go to the student support provision within the electronic resources, e.g. biochemistry support.

'**Review**' boxes and interactive GCSE review tests are provided to help you revise GCSE biology relevant to each topic.

At the end of each topic, as well as the '**check your notes**' activity for consolidation of each topic, there is an interactive '**Topic test**' box. This test will usually be set by your teacher/lecturer, and will help you to find out how much you have learned from the topic.

The key biological principle and all boxes linking to online resources are colour coded for each topic.

Thinking Bigger

At the end of each chapter of the book there is an opportunity to read and work with real-life research and writing about science. These sections will help you to expand your knowledge and develop your own research and writing techniques. The questions and tasks will help you to apply your knowledge to new contexts and to bring together different aspects of your learning from across the whole course.

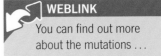

ACTIVITY

In **Student Activity 2.7** evaluate the evidence for …

WEBLINK

You can find out more about the mutations …

EXTENSION

You can read about the controversy …

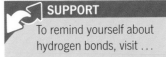

SUPPORT

To remind yourself about hydrogen bonds, visit …

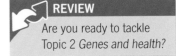

REVIEW

Are you ready to tackle Topic 2 *Genes and health*?

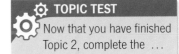

TOPIC TEST

Now that you have finished Topic 2, complete the …

Getting the most from your ActiveBook

Your ActiveBook is the perfect way to personalise your learning as you progress through your A level Biology course. You can:

● access your content online, anytime, anywhere

● use the inbuilt highlighting and annotation tools to personalise the content and make it really relevant to you

● search the content quickly.

Highlight tool

Use this to pick out key terms or topics so you are ready and prepared for revision.

Annotations tool

Use this to add your own notes, for example links to your wider reading, such as websites or other files. Or, make a note to remind yourself about work that you need to do.

Think critically and consider the issues

Analyse how scientists write

Reading material relevant to your course

Develop your own writing

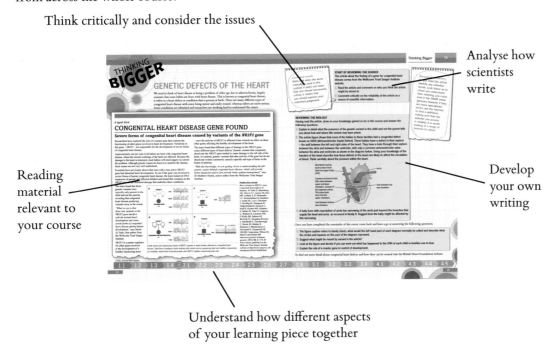

Understand how different aspects of your learning piece together

LIFESTYLE, HEALTH AND RISK

Why a topic called Lifestyle, health and risk?

Congratulations on making it this far! Not everyone who started life's journey has been so lucky. In the UK only about 80% of conceptions lead to live births, and about 4 in every 1000 newborn babies do not survive their first year of life (Figure 1.1). After celebrating your first birthday there seem to be fewer dangers. Fewer than 1 in every 1000 children die between the ages of 1 and 14 years old. All in all, life *is* a risky business.

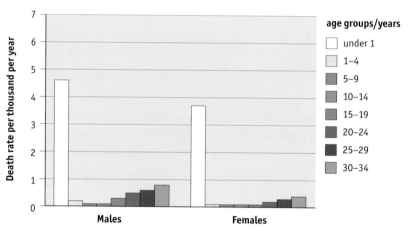

Figure 1.1 Death rates per 1000 population per year by age group and sex. Is life more risky for boys? *Source: England and Wales Office for National Statistics, 2012.*

Figure 1.2 Some activities are less obviously risky than others, but may still have hidden dangers.

In everything we do there is some risk. Normally we only think something is risky if there is the obvious potential for a harmful outcome. Snowboarding, parachute jumping and taking ecstasy are thought of as risky activities, but even crossing the road, jogging or sitting in the sun have risks, and many people take actions to reduce them (Figures 1.2 and 1.3).

Risks to health are often not as apparent as the risks facing someone making a parachute jump. People often do not realise that there are risks associated with lifestyle choices that they make. They underestimate the effect such choices might have on their health.

What we eat and drink, and the activities we take part in, all affect our health and well-being. Every day we make choices that may have short- and long-term consequences which we may be only vaguely aware of. What are the health risks that we are subjecting ourselves to? Will a cooked breakfast set us up for the day or will it put us on course for heart disease? Does the 10-minute walk to work really make a difference to our health?

Cardiovascular disease is the biggest killer in the UK, with around 1 in 3 people (32%) dying from diseases of the circulatory system. Does everyone have the same risk? Can we assess and reduce the risk to our health? Do we need to? Is our perception of risk at odds with reality?

In this topic you will read about Mark and Peter who have kindly agreed to share their experiences of cardiovascular disease. The topic will introduce the underlying biological concepts that will help you understand how cardiovascular diseases develop, and the ways of reducing the risk of developing these diseases.

OVERVIEW OF THE BIOLOGICAL PRINCIPLES COVERED IN THIS TOPIC

This topic will introduce the concept of risks to health. You will study the relative sizes of risks and how these are assessed. You will consider how we view different risks – our perception of risk. You will also look at how health risks may be affected by lifestyle choices and how risk factors for disease are determined.

Building on your GCSE knowledge of the circulatory system, you will study the heart and circulation and understand how these are affected by our choice of diet and activity.

You will look in some detail at the biochemistry of our food. This will give you a detailed understanding of some of the current thinking among doctors and other scientists about how our choice of foods can reduce the risks to our health.

REVIEW

Are you ready to tackle Topic 1 *Lifestyle, health and risk?*

Complete the GCSE review and GCSE review test before you start.

Figure 1.3 A UK male aged 15 to 19 is over three times more likely to have a fatal accident than a female of the same age. *Source: Department for Transport road accidents and safety annual report, 2012.*

Mark's story

On 28 July 1995 something momentous happened that changed my life ...

I was sitting in my bedroom playing on my computer when I started to feel dizzy with a slight headache. Standing, I lost all balance and was feeling very poorly. I think I can remember trying to get downstairs and into the kitchen before fainting. People say that unconscious people can still hear. I don't know if it's true but I can remember my dad phoning for a doctor and that was it. It took 5 minutes from me being an average 15-year-old to being in a coma.

Figure 1.4 Mark at 15.

I was rushed to Redditch Alexandra Hospital where they did some reaction tests on me. They asked my parents questions about my lifestyle (did I smoke, take drugs, etc.?). Failing to respond to any stimulus, I was transferred in an ambulance to Coventry Walsgrave Neurological Ward. Following CT and MRI scans on my brain it was concluded that I had suffered a stroke. My parents signed the consent form for me to have an operation lasting many hours. I was given about a 30% chance of survival.

They stopped the bleed by clipping the blood vessels that had burst with metal clips and removing the excess blood with a vacuum. I was then transferred to the intensive care unit to see if I would recover. Within a couple of days I was conscious and day by day I regained my sight, hearing and movement (although walking and speech were still distorted).

This is a true story. Mark had a stroke, one of the forms of cardiovascular disease. It is rare for someone as young as Mark to suffer a stroke. Why did it happen? Was he in a high-risk group?

Figure 1.5 The experience is not stopping Mark living life to the full. He is now married and works in IT.

Peter's story

I got the first indication of cardiovascular problems aged 23 when I was told that I had high blood pressure. I didn't really take much notice. My father had died at the age of 53 from a heart attack but as he was about four stone overweight, had a passion for fatty foods and smoked 60 full strength cigarettes a day, I didn't compare his condition to mine. I had a keen interest in sport, playing hockey and joining the athletics team at work. I was never overweight but I must admit that I probably drank too much at times and didn't bother too much about calories and cholesterol in food.

In 1981, I ran my first marathon at the age of 42 and subsequently did another five. All was going well I thought, until a routine medical showed my blood pressure reading to be very high at 240 over 140. The doctor could not believe that I was still walking around, let alone running, and sent me straight to my GP. Since then I have always taken tablets for high blood pressure and have also reviewed my diet.

I did continue running and completed the Great North Run at the age of 63. Thinking about doing the Great North Run again, I was running 8 miles a week and playing hockey. Then my eight-day holiday in Ireland became three days touring and twelve days in hospital.

At 2 o'clock in the morning I woke up with a terrific pain in my chest. I was sweating profusely and looking very pale. I had had a heart attack and within an hour I was in intensive care. At 5 am I had a second attack and the specialist inserted a temporary pacemaker to keep my heart rate up as it was dropping below 40.

After five days in intensive care I was transferred to the general ward for recuperation. I was told that it was possible that, had I not looked after myself, I might have had a heart attack much earlier in life.

On returning home I had an angiogram and was told that I needed a triple bypass operation. I have to say it was not pleasant, but I had decided that it was necessary and I would cope with anything that happened if it would get me back to a decent lifestyle. Well, the operation, a quadruple bypass, was a success and after eight days I was back home.

This is a true story. Why did it happen to Peter, who seemed to be so active and healthy?

 ACTIVITY
To find out what happened to Mark and Peter read their full stories in **Student Activity 1.1**.

Figure 1.6 Peter's active lifestyle did not prevent his heart attack but probably helped him to make a full recovery.

1.1 What is cardiovascular disease?

Deaths from cardiovascular disease

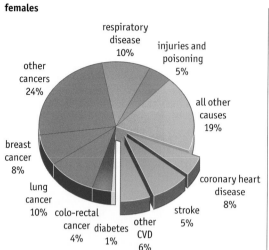

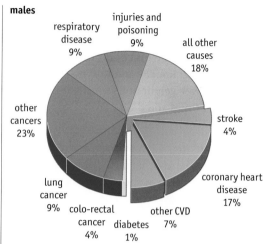

Figure 1.7 Premature deaths by cause in the UK in 2010 for females (left) and males (right). (Premature death is death under the age of 75 years.) One person dies of a heart attack in the UK every 7 minutes. *Reproduced with the kind permission of the British Heart Foundation.*

WEBLINK

To check out the most recent death rate figures for coronary heart disease see the National Statistics Office website and the British Heart Foundation website.

MATHS SUPPORT

Check why the data here in Figure 1.7 is presented as pie charts while the data in Figure 1.1 is in a histogram. See maths support 2 – presenting data graphs.

Cardiovascular diseases (CVDs) are diseases of the heart and circulation. They are the main cause of death in the UK, accounting for almost 180 000 deaths a year, and over 46 000 of these are premature deaths (Figure 1.7). Around one in three people in the UK die from cardiovascular diseases. The main forms of cardiovascular diseases are **coronary heart disease (CHD)**, as experienced by Peter, and **stroke**, as experienced by Mark.

Almost half of all deaths from cardiovascular diseases are from coronary heart disease (45%) and over a quarter are from stroke (28%). Coronary heart disease is the most common cause of death in the UK. About one in five men and one in ten women die from the disease.

KEY BIOLOGICAL PRINCIPLE: WHY HAVE A HEART AND CIRCULATION?

The heart and circulation have one primary purpose – to move substances around the body. In very small organisms such as unicellular creatures where distances are short, substances such as oxygen, carbon dioxide and digestive products move around the organism by diffusion. **Diffusion** is the movement of molecules or ions from a region of their high concentration to a region of their low concentration by relatively slow random movement of molecules. In unicellular organisms diffusion is usually fast enough to meet the organism's requirements.

Most complex multicellular organisms, however, are too large for diffusion to move substances around their bodies quickly enough. These organisms rely on a **mass transport system** to move substances efficiently over long distance by **mass flow**. All the particles in a liquid move in one direction through tubes due to difference in pressure. Animals usually have blood to carry vital substances around their bodies and a heart to pump it instead of relying on diffusion. In other words, they have a circulatory system. Some animals have more than one heart – the humble earthworm, for instance, has five.

Open circulatory systems

In insects and some other animal groups, blood circulates in large open spaces. A simple heart pumps blood out into cavities surrounding the animal's organs. Substances can diffuse between the blood and cells. When the heart muscle relaxes, blood is drawn from the cavity back into the heart through small, valved, openings along its length.

Closed circulatory systems

Many animals, including all vertebrates, have a closed circulatory system in which the blood is enclosed within tubes – blood vessels. This generates higher blood pressures as the blood is forced along fairly narrow channels instead of flowing into large cavities. This means the blood travels faster and so the blood system is more efficient at delivering substances around the body:

- The blood leaves the heart under pressure and flows along **arteries** and then **arterioles** (small arteries) to **capillaries**.

- There are extremely large numbers of capillaries. These come into close contact with most of the cells in the body where substances are exchanged between blood and cells.
- After passing along the capillaries, the blood returns to the heart by means of **venules** (small veins) and then **veins**.
- Valves ensure that blood flows only in one direction.

Animals with closed circulatory systems are generally larger in size and often more active than those with open systems.

Single circulatory systems

Animals with a closed circulatory system have either single circulation or double circulation. Fish, for example, have single circulation (Figure 1.8):

- The heart pumps deoxygenated blood to the gills.
- Gaseous exchange takes place in the gills; there is diffusion of carbon dioxide from the blood into the water that surrounds the gills, and diffusion of oxygen from this water into the blood within the gills.
- The blood leaving the gills then flows round the rest of the body before eventually returning to the heart.

Note that the blood flows through the heart once for each complete circuit of the body.

Double circulatory systems

Birds and mammals have double circulation:

- The right ventricle of the heart pumps deoxygenated blood to the lungs where it receives oxygen.
- The oxygenated blood then returns to the heart to be pumped a second time (by the left ventricle) out to the rest of the body.

This means that the blood flows through the heart twice for each complete circuit of the body. The heart gives the blood returning from the lungs an extra 'boost' that reduces the time it takes for the blood to circulate round the whole body. This allows birds and mammals to have a high metabolic rate, as oxygen and food substances required for metabolic processes can be delivered more rapidly to cells and meet the needs of the organism.

Q 1.1 Why do only small animals have an open circulatory system?

Q 1.2 What are the advantages of having a double circulatory system?

Q 1.3 Fish have two-chamber hearts and mammals have four-chamber hearts.

(a) Sketch what the three-chamber heart of an amphibian, such as a frog, might look like.

(b) What might be the major disadvantage of this three-chamber system?

> **✓ CHECKPOINT**
>
> **1.1** Make a bullet point summary which explains why many animals have a heart and circulation.

> **⚙ ACTIVITY**
>
> **Student Activity 1.2** demonstrates mass flow.

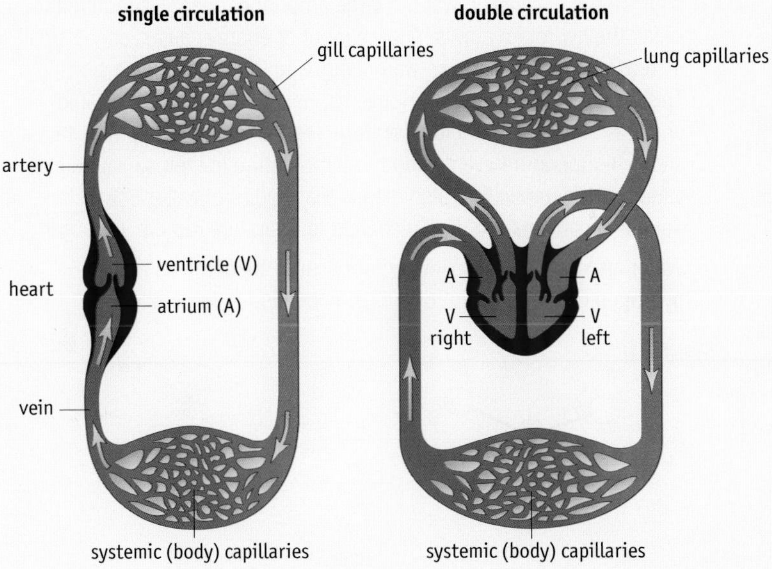

Figure 1.8 Fish have a single circulation. Birds and mammals have a double circulation.

How does the circulation work?

The transport medium

In the circulatory system a liquid and all the particles it contains are transported in one direction due to a difference in pressure in a process known as **mass flow**. In animals the transport medium is usually called blood. The fluid, plasma, is mainly water and contains dissolved substances such as digested food molecules (e.g. glucose), oxygen and carbon dioxide. Proteins, amino acids, salts, enzymes, hormones, antibodies and urea, the waste product from the breakdown of proteins, are just some of the other substances transported in the plasma. Cells are also carried in the blood: red blood cells, white blood cells and platelets. Blood is not only important in the transport of dissolved substances and cells, but also plays a vital role in regulation of body temperature, transferring energy around the body.

> **⚙ ACTIVITY**
>
> **Student Activity 1.3** lets you investigate some of the properties of water.

KEY BIOLOGICAL PRINCIPLE: PROPERTIES OF WATER THAT MAKE IT AN IDEAL TRANSPORT MEDIUM

Water, H_2O, is unusual among small molecules. It is a liquid at room temperature while most other small molecules, such as CO_2 and O_2, are gases. Water is a **polar molecule**; it has an unevenly distributed electrical charge. The two hydrogens are pushed towards each other forming a V-shaped molecule (Figure 1.9). The hydrogen end of the molecule is slightly positive and the oxygen end is slightly negative because the electrons are more concentrated at that end. Water is said to be a **dipole**. It is this polarity that accounts for many of its biologically important properties.

The slightly positively charged end of a water molecule is attracted to the slightly negative ends of surrounding water molecules. This **hydrogen bonding** holds the water molecules together and results in many of the properties of water including being liquid at room temperature.

Solvent properties

Many chemicals dissolve easily in water, due to their dipole nature, allowing vital biochemical reactions to occur in the cytoplasm of cells. Free to move around in an aqueous environment, the chemicals can react, often with water itself being involved in the reactions (for example in hydrolysis and condensation reactions, see page 31). The dissolved substances can also be transported around organisms, in animals via the blood and lymph systems, and in plants through the xylem and phloem.

Ionic substances, such as sodium chloride (NaCl), dissolve easily in water. In the case of sodium chloride, the negative Cl^- ions are attracted to the positive ends of the water molecules while the positive Na^+ ions are attracted to the negative ends of the water molecules. The chloride and sodium ions become hydrated in aqueous solution, they become surrounded by water molecules.

Polar molecules also dissolve easily in water. Their polar groups, for example the –OH group in sugars or the amine group, $-NH_2$, in amino acids, become surrounded by water and go into solution. Such polar substances are said to be **hydrophilic** – 'water-loving'.

Non-polar, **hydrophobic** substances, such as lipids, do not dissolve in water. To enable transport in blood, lipids combine with proteins to form lipoproteins.

Thermal properties

The specific heat capacity of water, the amount of energy in joules required to raise the temperature of $1\,cm^3$ ($1\,g$) of water by $1\,°C$, is very high. This is because in water a large amount of energy is required to break the hydrogen bonds. A large input of energy causes only a small increase in temperature, so water warms up and cools down slowly. This is extremely useful for organisms, helping them to avoid rapid changes in their internal temperature and enabling them to maintain a steady temperature even when the temperature in their surroundings varies considerably. This also means that bodies of water in which aquatic organisms live do not change temperature rapidly.

Water also has a high boiling point because there are so many hydrogen bonds and a lot of energy is needed to break them all.

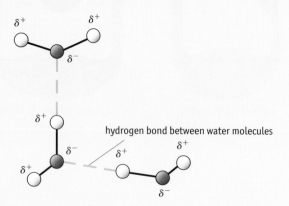

Figure 1.9 The polarity of the water molecules results in hydrogen bonds between them. (Oxygen atoms in red, hydrogen atoms in white.)

The structure of the heart

The heart is a double pump and is made of **cardiac muscle**. The right side of the heart receives deoxygenated blood from the body and pumps it to the lungs. The left side receives oxygenated blood from the lungs and pumps it to the body.

Study Figure 1.10 and locate the arteries carrying blood away from the heart and the veins returning blood to the heart.

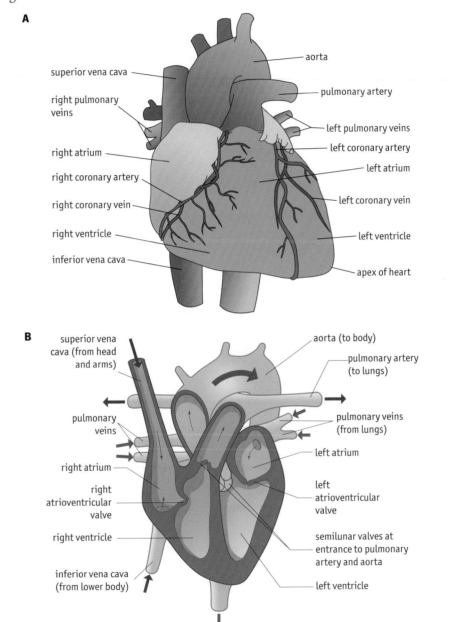

⚙ **ACTIVITY**

Student Activities 1.4 and **1.5** let you look in detail at the structure of a mammalian heart using either a dissection or a simulation.

Figure 1.10 A Diagrammatic external view of a human heart. **B** Diagrammatic cross-section of the human heart (ventral or front view).

The structure of blood vessels

Arteries and veins can easily be distinguished, as shown in Figure 1.11. The walls of both vessels contain **collagen**, a tough fibrous protein, which makes them strong and durable. They also contain elastic fibres that allow them to stretch and recoil. Smooth muscle cells in the walls allow them to constrict and dilate. The key differences between the arteries and veins are listed below.

Arteries:

● narrow lumen

● thicker walls

● more collagen, smooth muscle and elastic fibres

● no valves

Veins:

● wide lumen

● thinner walls

● less collagen and smooth muscle, fewer elastic fibres

● valves

A

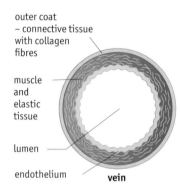

outer coat
– connective tissue with collagen fibres

muscle and elastic tissue

lumen

endothelium

artery

outer coat
– connective tissue with collagen fibres

muscle and elastic tissue

lumen

endothelium

vein

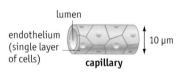

lumen

endothelium (single layer of cells)

 10 µm

capillary

B

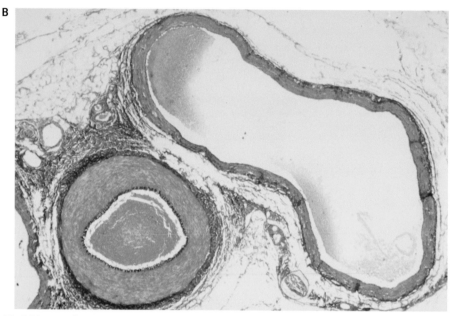

Figure 1.11 A Diagram of an artery, a vein and a capillary. The endothelium that lines the blood vessels is made up of epithelial cells (see page 59). **B** Photomicrograph of an artery (left) and vein (right) surrounded by connective tissue.

Q 1.4 **(a)** A student calibrating her eyepiece graticule found 5 units measured 3.5 units on the stage micrometer, which are each 1 mm in length. Work out the length of one eyepiece graticule unit in µm.

(b) Using the same eyepiece graticule (epg). The width of the artery wall in the photomicrograph in Figure 1.11B measured 0.2 epg units at its widest point. **(i)** what was the width in µm and **(ii)** what is the magnification of the photograph?

The capillaries that join the small arteries (arterioles) and small veins (venules) are very narrow, about 10 µm in diameter, with walls that are only one cell thick.

These features can be directly related to the functions of the blood vessels, as described below.

How does blood move through the vessels?

Every time the heart contracts (**systole**), blood is forced into arteries and their elastic walls stretch to accommodate the blood. The thick artery walls can withstand the high pressure generated as the blood is forced against the walls. During **diastole** (relaxation of the heart), the elasticity of the artery walls causes them to recoil behind the blood, helping to push the blood forward and smoothing blood flow. The blood moves along the length of the artery as each section in series stretches and recoils in this way. The pulsing flow of blood through the arteries can be felt anywhere an artery passes over a bone close to the skin.

⚙ ACTIVITY

Student Activity 1.6 lets you investigate how the structure of blood vessels relates to their function. You will also learn how to measure using an eyepiece graticule.

By the time the blood reaches the smaller arteries and capillaries there is a steady flow of blood. Blood flows more slowly in the capillaries due to their narrow lumens causing more of the blood to be slowed down by friction against the capillary wall. This slower steady flow allows exchange between the blood and the surrounding cells through the one-cell-thick capillary walls. The network of capillaries that lies close to every cell ensures that there is rapid diffusion between the blood and surrounding cells.

The heart has a less direct effect on the flow of blood through the veins. Blood flows steadily and without pulses in veins where it is under relatively low pressure. In the veins blood flow is assisted by the contraction of skeletal muscles during the movement of limbs and breathing. Low pressure developed in the thorax (chest cavity) when breathing in also helps draw blood back into the heart from the veins. Backflow is prevented by semilunar valves within the veins (Figure 1.12).

CHECKPOINT

1.2 Identify the key structures of an artery, a vein and a capillary, and in each case explain how the structure is related to the function of the vessel.

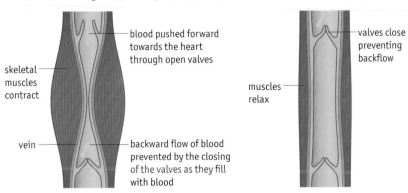

Figure 1.12 Valves in the veins prevent the backflow of blood.

Q 1.5 List the features shown in Figure 1.11A that enable the artery to withstand high pressure and then recoil to maintain a steady flow of blood.

Since the heart is a muscle it needs a constant supply of fresh blood carrying oxygen and glucose for aerobic respiration. You might think that receiving a blood supply would never be a problem for the heart. However, the heart muscle does not obtain oxygen and nutrients from the blood inside its pumping chambers due to the large diffusion distances involved. Instead, the heart muscle is supplied with blood through its own coronary circulation; two vessels called the **coronary arteries**, a network of capillaries, and two coronary veins. You can see the coronary arteries and coronary veins on the surface of the heart in Figure 1.10A.

ACTIVITY

Student Activity 1.7 lets you complete William Harvey's experiment that originally demonstrated one-way valves in veins.

How the heart works

Give a tennis ball a good, hard squeeze. You are using about the same amount of force that your heart uses in a single contraction to pump blood out to the body. Even when you are at rest, the muscles of your heart work hard – weight for weight, harder than the leg muscles of a person running.

The chambers of the heart alternately contract (systole) and relax (diastole) in a rhythmic cycle. One complete sequence of filling and pumping blood is called a **cardiac cycle**, or heartbeat. During systole, cardiac muscle contracts and the heart pumps blood out through the aorta and pulmonary arteries. During diastole, cardiac muscle relaxes and the heart fills with blood.

The cardiac cycle can be simplified into three phases: atrial systole, ventricular systole and diastole. The events that occur during each of the stages are shown in Figure 1.13.

Phase 1: Atrial systole

Blood returns to the heart due to the action of skeletal and muscles involved in breathing as you move and breathe. Blood under low pressure flows into the **left** and **right atria** from the pulmonary veins and vena cava. As the atria fill, the increasing pressure of blood against the **atrioventricular valves** pushes them open and blood begins to leak into the **ventricles**. The atria walls contract forcing more blood into the ventricles. This contraction of the atria is known as **atrial systole**.

Phase 2: Ventricular systole

After a *slight* delay, atrial systole is followed by **ventricular systole**. The ventricles contract from the base of the heart upwards, increasing the pressure in the ventricles. The pressure forces open the semilunar valves and pushes blood up and out through the pulmonary arteries and aorta. The pressure of blood against the atrioventricular valves closes them and prevents blood flowing backwards into the atria.

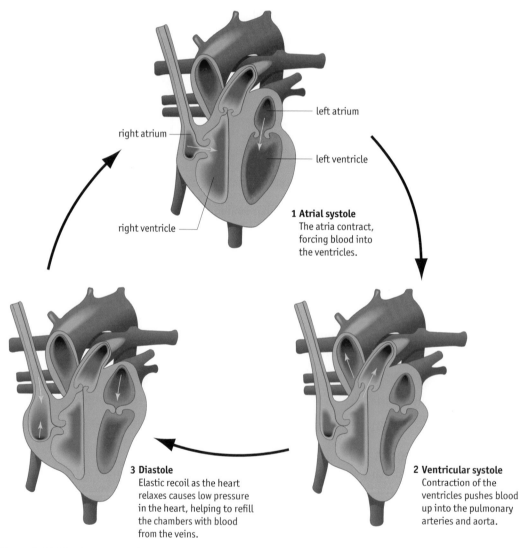

right atrium

left atrium

left ventricle

right ventricle

1 Atrial systole
The atria contract, forcing blood into the ventricles.

3 Diastole
Elastic recoil as the heart relaxes causes low pressure in the heart, helping to refill the chambers with blood from the veins.

2 Ventricular systole
Contraction of the ventricles pushes blood up into the pulmonary arteries and aorta.

Figure 1.13 The three stages of the cardiac cycle. At each stage blood moves from higher to lower pressure.

Phase 3: Cardiac diastole

The atria and ventricles then relax during **cardiac diastole**. Elastic recoil of the relaxing heart walls lowers pressure in the atria and ventricles. Blood under higher pressure in the pulmonary arteries and aorta is drawn back towards the ventricles, closing the **semilunar valves** and preventing further backflow into the ventricles. The coronary arteries fill during diastole. Low pressure in the atria helps draw blood into the heart from the veins.

> **✓ CHECKPOINT**
> **1.3** Make a flowchart which summarises the events in the cardiac cycle.

Q 1.6 When the heart relaxes in cardiac diastole you might expect blood to move from the arteries back into the ventricles due to the elastic recoil of the heart and the action of gravity if you are standing or sitting upright. How is this prevented?

⚙ **ACTIVITY**
Student Activity 1.8 lets you test your knowledge of the cardiac cycle.

Pressure changes and valves determine the flow of blood in the cardiac cycle. At each stage in the cycle blood moves from high pressure to low pressure. Figure 1.14 shows changes in pressure in the left side of the heart during the cardiac cycle. The same sequence occurs in the right side of the heart but the maximum pressure in the right ventricle is only 30 mm Hg. The diagram also shows how the closing of the valves causes the sounds that we recognise as a heartbeat. The first sound ('lub') is caused by the closing of the atrioventricular valves and the second ('dub') by the closing of the semilunar valves.

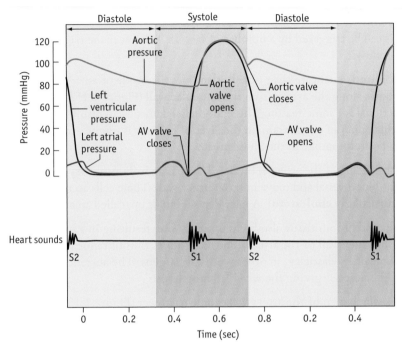

Figure 1.14 Pressure changes in the left side of the heart during the cardiac cycle. The differences in pressure determine the movement of blood and the opening and closing of the valves, and therefore maintain the flow of blood in one direction through the heart. The aortic valve is the semilunar valve in the aorta. Complete question 1.7 to make sure that you really understand what this diagram is showing you.

Q 1.7 (a) Using your knowledge of the cardiac cycle and information from the graph explain what causes:
- **(i)** the rise in both atrial and ventricle pressure at about 0.3 s
- **(ii)** the atrioventricular valve to close at about 0.45 seconds
- **(iii)** the semilunar valve (aortic) to open at about 0.5 seconds
- **(iv)** the rise in aorta pressure after the semilunar valve opens
- **(v)** the closing of the semilunar valve (aortic) at about 0.75 seconds.

(b) Decide what state the valves will be in, open or closed, for each of the pressure gradients shown.

Valves	Pressure gradient
(i) Atrioventricular valves	Atrium pressure > ventricle pressure
(ii) Semilunar valves	Ventricle pressure < aorta pressure

(c) Sketch a graph to show the pressure changes that would occur on the right side of the heart during a single cardiac cycle.

(d) Work out the heart rate for the cardiac cycle shown in Figure 1.14.

What is atherosclerosis?

Atherosclerosis is the disease process that leads to coronary heart disease and strokes. In atherosclerosis fatty deposits can either block an artery directly or increase its chance of being blocked by a blood clot (**thrombosis**). The blood supply can be blocked completely. If it is not restored very quickly, the affected cells are permanently damaged. In the coronary arteries this results in a heart attack (**myocardial infarction**). In the arteries supplying the brain it results in a **stroke**. The supply of blood to the brain is restricted or blocked, causing damage or death to cells in the brain. Narrowing of arteries to the legs can result in tissue death and gangrene (decay). An artery can burst where blood builds up behind an artery that has been narrowed as a result of atherosclerosis (see page 17 Did You Know?).

What happens in atherosclerosis?

Atherosclerosis can be triggered by a number of factors. Whatever the trigger, this is the course of events that follows:

1 The **endothelium**, a delicate layer of cells that lines the inside of an artery and separates the blood that flows along the artery from the muscular wall (Figure 1.15A), becomes damaged and dysfunctional for some reason. This endothelial damage can result from high blood pressure, which puts an extra strain on the layer of cells, or it might occur due to some of the toxins from cigarette smoke in the bloodstream.

2 Once the inner lining of the artery is breached there is an **inflammatory response**. White blood cells leave the blood vessel and move into the artery wall. These cells accumulate chemicals from the blood, particularly **cholesterol**. A fatty deposit builds up, called an **atheroma**.

3 Calcium salts and fibrous tissue also build up at the site, resulting in a hard swelling called a **plaque** on the inner wall of the artery. The build-up of fibrous tissue means that the artery wall loses some of its elasticity; in other words, it hardens. The ancient Greek word for 'hardening' is 'sclerosis', giving the word 'atherosclerosis'.

4 Plaques cause the lumen of the artery to become narrower (Figure 1.15B). This makes it more difficult for the heart to pump blood around the body and can lead to a rise in blood pressure. Now there is a dangerous **positive feedback** building up. Plaques lead to raised blood pressure and raised blood pressure makes it more likely that further plaques will form, as damage to endothelial tissue in other areas becomes more likely.

The person may be unaware of any problem at this stage, but if the arteries become very narrow or completely blocked they cannot supply enough blood to bring oxygen and nutrients to the tissues. The tissues can no longer function normally and symptoms will soon start to show.

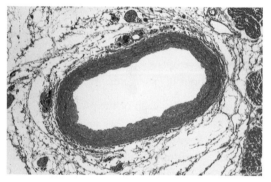

Figure 1.15 A Photomicrograph of a normal, healthy coronary artery showing no thickening of the arterial wall. The lumen is large. Magnification ×15.

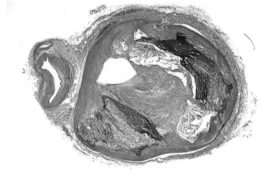

Figure 1.15 B Photomicrograph of a diseased coronary artery showing narrowing of the lumen due to atheroma deposits and build-up of atherosclerotic plaque. Magnification ×15.

Why do only arteries get atherosclerosis?

The fast-flowing blood in arteries is under high pressure so there is a significant chance of damage to the walls. The low pressure in the veins means that there is less risk of damage to the walls.

Why does the blood clot in arteries?

Blood clotting

Rapid blood clotting is vital when a blood vessel is damaged. The blood clot seals the break in the blood vessel and limits blood loss and prevents entry of pathogens through any open wounds. When **platelets**, a type of blood cell without a nucleus, come into contact with the damaged vessel wall they change from flattened discs to spheres with long thin projections (Figure 1.16). Their cell surfaces change, causing them to stick to the exposed collagen in the wall and to each other to form a temporary platelet plug. They also release substances that activate more platelets.

The direct contact of blood with collagen within the damaged blood vessel wall also triggers a complex series of chemical changes in the blood (Figure 1.17). A **cascade** of changes results in the formation of a blood clot (Figures 1.17 and 1.18).

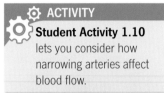

ACTIVITY
Student Activity 1.9 lets you summarise the steps in development of atherosclerosis and clot formation.

ACTIVITY
Student Activity 1.10 lets you consider how narrowing arteries affect blood flow.

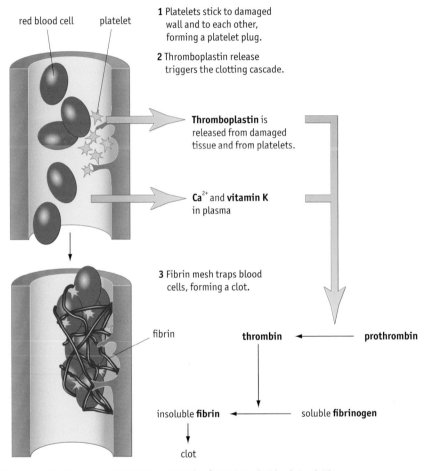

1 Platelets stick to damaged wall and to each other, forming a platelet plug.

2 Thromboplastin release triggers the clotting cascade.

red blood cell   platelet

Thromboplastin is released from damaged tissue and from platelets.

Ca^{2+} and **vitamin K** in plasma

3 Fibrin mesh traps blood cells, forming a clot.

fibrin

thrombin ← prothrombin

insoluble **fibrin** ← soluble **fibrinogen**

clot

Figure 1.17 Damage to the vessel wall triggers a cascade of reactions that leads to clotting.

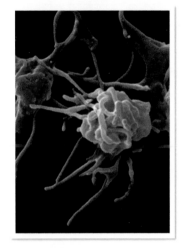

Figure 1.16 Electron micrograph showing activated platelets. Magnification ×6000.

The clotting cascade

1 Platelets and damaged tissue release a protein called **thromboplastin**.

2 Thromboplastin activates an enzyme that catalyses the conversion of the protein **prothrombin** into an enzyme called **thrombin**. A number of other protein factors, vitamin K and calcium ions must be present in the blood plasma for this conversion to happen.

3 Thrombin then catalyses the conversion of the soluble plasma protein, **fibrinogen**, into the insoluble protein **fibrin**.

4 A mesh of fibrin forms that traps more platelets and red blood cells to form a clot.

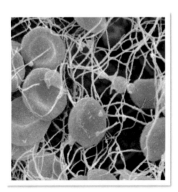

Figure 1.18 False-colour scanning electron micrograph showing red blood cells and platelets trapped in the yellow mesh of fibrin.

What happens inside arteries to cause blood clotting?

Usually blood does not clot inside blood vessels. Platelets do not stick to the endothelium (inner lining) of blood vessels. It is very smooth and has substances on its surface that repel the platelets. However, if there is atherosclerosis and the endothelium is damaged, the platelets come into contact with the damaged surface and any exposed collagen. The clotting cascade will be triggered within the vessel resulting in a clot as shown in Figure 1.19.

The consequences of atherosclerosis

Coronary heart disease

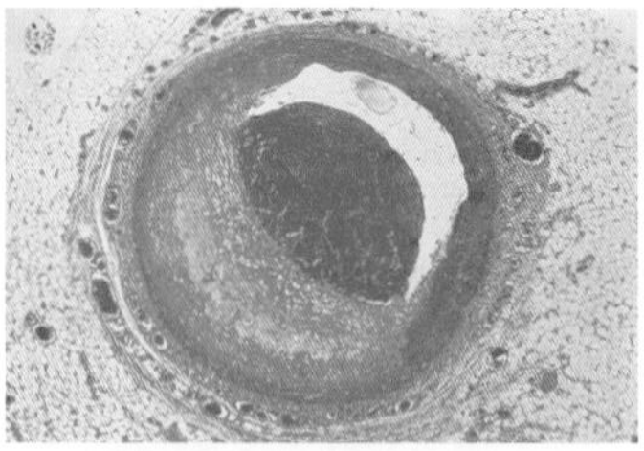

Narrowing of the coronary arteries limits the amount of oxygen-rich blood reaching the heart muscle. The result may be a chest pain called **angina**. Angina is usually experienced during exertion when the cardiac muscle is working harder and needs to respire more. Because the heart muscle lacks oxygen, it is forced to respire **anaerobically**. It is thought that this results in chemical changes which trigger pain, but the detailed mechanism is still not known. Usually these symptoms will ease with rest.

If a fatty plaque in the coronary arteries ruptures, collagen is exposed which leads to rapid clot formation. The blood supply to the heart may be blocked completely. The heart muscle supplied by these arteries does not receive any blood, so it is said to be **ischaemic** (without blood). If the affected muscle cells are starved of oxygen for long they will be permanently damaged. This is what we call a **heart attack** or **myocardial infarction**. If the zone of dead cells occupies only a small area of tissue the heart attack is less likely to prove fatal (see Figure 1.20).

Figure 1.19 Photomicrograph of a diseased coronary artery showing narrowing and a blood clot.

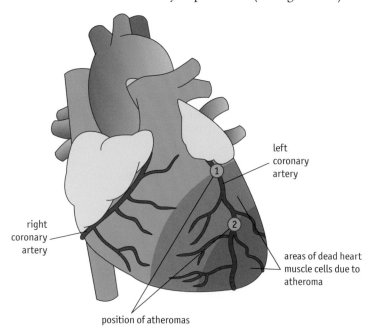

Figure 1.20 The seriousness of a heart attack is determined by the position of the blockage in the coronary artery. Blockage at position one is more likely to be fatal than a blockage at position two. Blockage at the position further along the coronary artery is less likely to be fatal.

Stroke

If the supply of blood to the brain is only briefly interrupted then a mini-stroke may occur. A mini-stroke has all the symptoms of a full stroke but the effects last for only a short period, and full recovery can happen quite quickly. However, a mini-stroke is a warning of problems with blood supply to the brain that could result in a full stroke in the future.

If a blood clot blocks one of the arteries leading to the brain, a full **stroke** will result. If brain cells are starved of oxygen for more than a few minutes they will be permanently damaged, and it can be fatal.

The symptoms of cardiovascular disease

Coronary heart disease

Shortness of breath and angina are often the first signs of coronary heart disease. The main symptom of angina is intense pain, ache or a feeling of constriction and discomfort in the chest, or in the left arm and shoulder. Other symptoms are unfortunately very similar to those of severe indigestion and include a feeling of heaviness, tightness, pain, burning and pressure – usually behind the breastbone, but sometimes in the jaw, arm or neck. Women may not have chest pain but experience unusual fatigue, shortness of breath and indigestion-like symptoms.

Sometimes coronary heart disease causes the heart to beat irregularly. This is known as **arrhythmia** and can itself lead to heart failure. Arrhythmia can be important in the diagnosis of coronary heart disease.

Stroke

The effects of a stroke will vary depending on the type of stroke, where in the brain the problem has occurred, and the extent of the damage. The more extensive the damage, the more severe the stroke and the lower the chance of full recovery. The symptoms normally appear very suddenly and include:

numbness
dizziness
confusion
slurred speech
blurred or lost vision, often only in one eye.

Visible signs often include paralysis on one side of the body with a drooping arm, leg or eyelid, or a dribbling mouth. The *right* side of the brain controls the *left* side of the body, and vice versa, therefore the paralysis occurs on the opposite side of the body to where the stroke occurred.

Aneurysms

If part of an artery has narrowed and become less flexible, blood can build up behind it. The artery bulges as it fills with blood and an **aneurysm** forms. An atherosclerotic aneurysm of the aorta is shown in Figure 1.21.

What will eventually happen as the bulge enlarges and the walls of the aorta are stretched thin? Aortic aneurysms are likely to rupture when they reach about 6–7 cm in diameter. The resulting blood loss and shock can be fatal. Fortunately, earlier signs of pain may prompt a visit to the doctor. The bulge can often be felt in a physical examination or seen with ultrasound examination and it may be possible to surgically replace the damaged artery with a section of artificial artery.

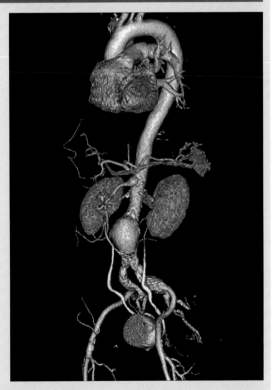

Figure 1.21 An aneurysm in the aorta below the kidneys. If an aneurysm ruptures it can be fatal.

EXTENSION

Read **Student Extension 1.1** to find out how you may be able to save someone's life by carrying out cardiopulmonary resuscitation.

EXTENSION

There are several tests used to diagnose cardiovascular disease that can be requested by doctors and you can read more details of these tests in **Student Extension 1.2**.

1.2 Who is at risk of cardiovascular disease?

Probability and risk

What do we mean by risk?

Risk is defined as 'the probability of occurrence of some unwanted event or outcome'. It is usually in the context of hazards, that is, anything that can potentially cause harm, such as the chance of contracting lung cancer if you smoke. Probability has a precise mathematical meaning and can be calculated to give a numerical value for the size of the risk. Do not panic – the maths is simple!

Taking a risk is a bit like throwing a die (singular of 'dice'). You can calculate the chance that you will have an accident or succumb to a disease (or throw a six). You will not *necessarily* suffer the accident or illness, but by looking at past circumstances of people who have taken the same risk, you can estimate the chance that you will suffer the same fate to a reasonable degree of accuracy.

Working out probabilities

There are six faces on a standard die. Only one face has six dots, so the chance of throwing a six is 1 in 6 (provided the die is not loaded). Scientists tend to express '1 in 6' as a decimal: 0.166 666 recurring (about 0.17). In other words, each time you throw a standard die, you have about a 0.17 or 17% chance of throwing a one, about a 17% chance of throwing a two, and so on.

When measuring risk you must always quote a time period for the risk. Here you have a 17% chance of throwing a one with each throw of the die.

In a Year 5 class of 30 pupils, six children caught head lice in one year. The risk of catching head lice in this class was therefore 6 in 30, or 1 in 5, giving a probability of 0.2 or 20% in a year.

 (a) In 2011 there were 727 724 recorded births in England and Wales. Of these, 3811 were stillbirths. Work out the chance of having a stillbirth in 2011.

(b) In 2012 there were 733 232 recorded births. There were 6.6% fewer still births than in 2011. Calculate how many stillbirths there were in 2012, and then calculate the probability of having a stillbirth in 2012.

Estimating risks to health

In 2010, 19 900 people in the UK died due to injuries or poisoning. The total UK population at the time was 62 262 000, so we can calculate the average risk in a year of someone in the UK dying from injuries or poisoning as:

19 900 in 62 262 00

or 1 in $\dfrac{62\,262\,00}{19\,900}$

= 1 in 3129

= $\dfrac{1}{3129}$

= 0.000 32 or 0.032%

Another way of working this out is as:

$\dfrac{19\,900}{62\,262\,00}$ = 0.000 32

However, when calculating a probability in relation to health, most people would find, for example, 1 in 3129 more meaningful than 0.000 32 or 0.032%.

SUPPORT

For help with working out probabilities look at the maths support sheet 8 – probability.

Assuming the proportion of people that die from injuries or poisoning remains much the same each year, this calculation gives an estimate of the risk for any year.

If we calculated the risk of any one of us developing lung cancer in our lifetime we would find a probability of 1 in about 1600. However, because lung cancer is much more likely if you smoke, the risk for smokers is far greater. When looking at calculated risk values you need to think about exposure to the hazard.

Q 1.9 Look at the causes of death listed below and put them in order, from the most likely to the least likely. You could also have a go at estimating the percentage probability of someone in the UK dying from each cause during a year.

- accidental poisoning
- heart disease
- injury purposely inflicted by another person
- lightning
- lung cancer
- railway accidents
- road accidents

Did you get it right?

People frequently get it wrong, underestimating or overestimating risk. We can say that there is about a 1 in 1700 risk of each of us dying from lung cancer in any one year, a 1 in 100 000 risk of our being murdered in the next 12 months, and a 1 in 1.7 million risk of our being involved in a fatal rail accident in a year. However, recent work on risk has concentrated not so much on numbers such as these but on the perception of risk.

Perception of risk

The significance of the perception of risk can be illustrated by decisions about eligibility for blood donation made by the American Red Cross, which provides about half of the USA's blood supplies. In 2001, they decided to ban all blood donations from anyone who has spent six months or more in any European country since 1980. They now ban blood donations from anyone who spent three or more months in the UK between 1980 and 1996, and from anyone spending five years or more in Europe since 1980. Their reason is the risk of transmitting variant Creutzfeldt–Jakob disease (vCJD) through blood transfusion. There is a chance of this happening. In the UK there have been a total of 174 cases of this fatal condition, which causes brain damage, with four cases associated with blood transfusions between 1996 and 1999. However, in the UK, only individuals who have received a blood transfusion since 1980 are ineligible to donate blood. As the USA is short of blood for blood transfusions, it is possible that more people may have died as a result of these 'safety precautions' than would have been the case without them.

So why did America ban European blood donations? The likely reason was public perceptions of the risk of contracting vCJD. People will *overestimate* the risk of something happening if the risk is:

- involuntary (not under their control)
- not natural
- unfamiliar
- dreaded
- unfair
- very small.

If you look at this list you should be able to see why people may greatly overestimate some risks (such as the chances of contracting vCJD from blood transfusions) while underestimating others (such as the dangers of driving slightly faster than the speed limit or playing on a frozen lake).

ACTIVITY

Student Activity 1.11 asks you to estimate risks for a range of diseases using National Office for Statistics data.

WEBLINK

You can find out more about vCJD, the human form of bovine spongiform encephalopathy (BSE), by visiting the World Health Organisation website or the National CJD Research and Surveillance Unit website.

Nowadays many risk experts argue that perceptions of risk are what really drive people's behaviour. Consider what happened when it became compulsory in the UK to use seat belts for children in the rear seats of cars (Figure 1.22). The number of children killed and injured *increased*. How could this be? John Adams, an academic at University College London, argues that this is because the parents driving felt safer once their children were wearing seat belts and so drove slightly less carefully. Unfortunately, this change in their driving behaviour was more than enough to compensate for any extra protection provided by the seat belts.

There is a tendency to overestimate the risks of sudden imposed dangers where the consequences are severe, and to *underestimate* a risk if it has an effect in the long-term future, even if that effect is severe, for example, the health risks associated with smoking or poor diet.

A useful distinction is sometimes made between risk and uncertainty. When we lack the data to estimate a risk precisely we are *uncertain* about the risk. For example, we are uncertain about the environmental consequences of many chemicals.

Q 1.10 **(a)** In a school of 1300 students, in one term 10 students contracted verrucas from the school pool. In a letter to parents the head teacher said there was a less than 1% chance of any child catching a verruca in any term. Was the figure she quoted correct and what assumptions had she made in making this statement?

(b) In 2013, 208 755 cases of chlamydia were reported in England, with 43 386 of these cases being reported in London. One newspaper wanted to write a front page headline claiming that there was a higher risk of contracting this sexually transmitted infection in the capital compared with the rest of the country. Would they have been correct? Support your answer with calculated risk values. The population of England in 2013 was 53.5 million; the population of London was 8.3 million.

Different types of risk factor

In the UK the estimated risk of any one of us having fatal heart disease in any one year is about 1 in 600, compared with 1 in 1050 for a fatal stroke. However, these probabilities use figures for the whole population, giving averages which make the simplistic assumption that everyone has the same chance of having cardiovascular disease. This is obviously not the case.

The averages take no account of any risk factors – things that increase the chance of the harmful outcome. When assessing an individual's risk of bad health, all the contributing risk factors need to be established.

There are many different factors that contribute to health risks, for example:

- age
- heredity
- physical environment
- social environment
- lifestyle and behaviour choices.

Identifying risk factors – correlation and causation

To determine what the risk factors are for a particular disease, scientists look for **correlations** between potential risk factors and the occurrence of the disease. There is a correlation between two variables when a change in one variable is accompanied by a change in the other.

Two variables are *positively correlated* when an increase in one is accompanied by an increase in the other (Figure 1.23A). For example, the length of a TV programme and the percentage of the class asleep might be positively correlated. There is a positive correlation between the number of cigarettes smoked over a lifetime and the chance of developing cardiovascular disease. If the values of one variable decrease while the other increases, there is a *negative correlation* (Figure 1.23B).

Figure 1.22 Some research suggests that young children who wear rear seat belts are more likely to die in an accident than those who don't, but this may be explained by parents' driving habits. Health risks are greatly affected by human behaviour.

✓ CHECKPOINT

1.4 List the circumstances that make people more likely to **a** underestimate and **b** overestimate the risk of an event happening. Suggest an example for each situation.

SUPPORT

To find out more about correlations and how to statistically test whether there is a correlation between two variables see maths support sheet 12 – Spearman's rank correlation.

Large amounts of data are needed to ensure that the correlation is statistically significant; in other words, not just an apparent correlation due to chance.

It is important to realise that a correlation between two variables does not necessarily mean that the variables are causally linked. Two variables are *causally linked* when a change in one is responsible for a change in the other. It is easy to think of variables that are correlated where there is no **causation**. For example, worldwide, speaking English as your first language correlates quite well with having a greater-than-average life expectancy. This, though, is simply because countries like the USA, UK, Australia and Canada have a higher-than-average standard of living. It is this that causes increased life expectancy through better nutrition, medical care and so on, rather than the language spoken.

It is because of this logical gap between correlation and causation that scientists try, whenever they can, to carry out experiments in which they can control variables, to see if altering one variable really does have the predicted effect. To do this, scientists often set up a **null hypothesis**. They assume for the sake of argument that there will be no difference between an experimental group and a control group, and then test this hypothesis using statistical analysis.

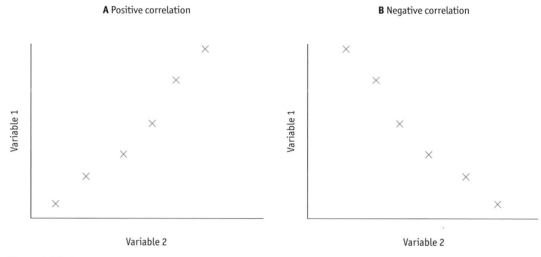

A Positive correlation   **B** Negative correlation

Figure 1.23 A When an increase in one variable is accompanied by an increase in the other, there is a positive correlation, giving a scattergram rising from left to right. **B** With a negative correlation, one set of data increases while the other falls, resulting in a graph going down from left to right.

Q 1.11 Strong correlations have been reported between the following pairs of variables. In each case, decide if it is a positive or negative correlation and if there is likely to be a causal link between the variables or not. Suggest a possible reason for the correlation.

(a) shark attacks and ice cream sales

(b) children's foot sizes and their spelling abilities

(c) lung cancer and smoking

(d) number of alcoholic drinks consumed and manual dexterity.

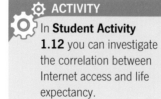

⚙ **ACTIVITY**
In **Student Activity**
1.12 you can investigate the correlation between Internet access and life expectancy.

1.3 Risk factors for cardiovascular disease

Identifying risk factors for CVD

Large-scale studies have been undertaken to find the risk factors for many common diseases, including cardiovascular disease. Epidemiologists, scientists who study patterns in the occurrence of disease, look for correlations between a disease and specific risk factors.

Two commonly used designs for this type of study are cohort studies and case-control studies.

Cohort studies

Cohort studies follow a large group of people over time to see who develops the disease and who does not (Figure 1.24). These types of studies are prospective; at the start of the study none of the participants have the disease. Researchers are interested in what happens to them in the future. During the study people's exposure to suspected risk factors and whether they develop the disease is recorded so any correlations between the risk factors and disease development can be identified. It may take a long time for the condition to develop so these studies can take years and be very expensive.

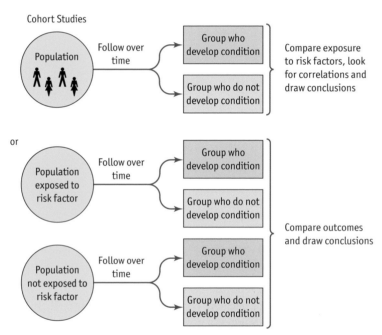

Figure 1.24 In cohort studies, risk factors experienced by those who develop the disease are compared with the risk factors of those who do not develop the disease to identify any correlations.

Cohort studies for CVD

The first major cohort study into CVD started in 1948. At the time, little was known about the causes of heart disease and stroke. The aim of the Framingham Heart Study was to identify the factors that contribute to the development of the disease. A random sample of 5209 men and women between the ages of 30 and 62 from the town of Framingham, Massachusetts, was recruited for the study. At the time of recruitment they had no symptoms of cardiovascular disease. In 1971, 5124 of the participants' adult children joined the study. A third generation was recruited to the study in 2002, adding a further 4095 individuals to the study, which continues to the present day.

Every two years the participants are asked to provide a detailed medical history, undergo a physical examination and tests, and answer questions about their lifestyle. The data are used to look for common features that contribute to the development of CVD. High blood pressure, high blood cholesterol, smoking, obesity, diabetes and physical inactivity were all identified as major CVD risk factors as a result of this study.

Other studies have confirmed these findings. The World Health Organization MONICA study (MONItoring trends and determinants in CArdiovascular disease), involving over 10 million people aged between 25–64 years old in 21 countries over 10 years, confirmed the link between several of these factors and increased occurrence of the disease. Although the study was completed in the late 1990's, the data is still being used for analysis in studies today.

Case-control studies

In a case-control study, a group of people with a disease (cases) are compared with a control group of individuals who do not have the disease (Figure 1.25). Information is collected about the risk factors that they have been exposed to in the past, allowing factors that may have contributed to development of the disease to be identified. These type of studies are retrospective.

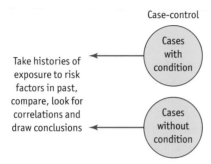

Figure 1.25 In case-control studies, risk factors experienced by those who have the disease are compared with the risk factors experienced by those who do not have the disease to identify any correlations.

The control group should be representative of the population from which the case group was drawn. Sometimes controls are individually matched to cases; known disease-risk factors, such as age and sex, are then similar in each case and control pair. This allows scientists to investigate the potential role of unknown risk factors. It should be noted that factors used to match the cases and controls cannot be investigated within the study, so it is important not to match any variables which could potentially turn out to be risk factors.

One of the first case-control studies was conducted in the 1950s by two British scientists, Richard Doll and Austin Bradford Hill, to determine whether there was a link between smoking and lung cancer. A group of hospital patients with lung cancer was compared with a second group who did not have cancer. The data indicated a correlation between smoking and lung cancer.

 Q 1.12 Did the study by Doll and Hill prove that smoking caused lung cancer? Give a reason for your answer.

Case-control studies and CVD

Put the terms case-control and coronary heart disease into any search engine and you will find numerous case-control studies investigating the risk factors for CVD, including whether factors such as passive smoking or a siesta can increase your risk.

One global case control study, the INTERHEART study, screened all patients with a first heart attack admitted to 262 participating hospitals in 52 countries. A total of 15 152 cases and 14 820 controls, matched by age and gender, were included in the study. The researchers' original hypothesis was that risk factors for cardiovascular disease differ between people of varying ethnic and geographic origin. However, the study results, published in *The Lancet* in 2004, concluded that nine risk factors accounted for over 90% of the risk and they are the same for men and women in almost every geographic region, and for every racial/ethnic group.

Features of a good study

To identify correlations between risk factors and disease, studies need to be carefully designed. Recording a higher rate of heart disease in 50 people who drink more alcohol than the recommended amount compared with 50 people who drink less than the recommended amount, supports the suggestion that excess alcohol consumption increases the risk of developing heart disease. However, the group who drink more alcohol might also smoke more, do less exercise and eat a fatty diet. Any of these factors could be linked to developing the disease. A well-designed study tries to overcome these problems. When designing an epidemiological study there are some key points to consider.

Clear aim

A well-designed study should include a clearly stated hypothesis or aim. The design of the study must be appropriate to the stated hypothesis or aim and produce results that are valid and reliable.

Representative sample

A representative sample must be selected from the wider population that the study's conclusions will be applied to. Selection bias occurs when those who participate in a study are not representative of the target population. For example, if a study aiming to look at the prevalence of disease in a community only sent out questionnaires to people on the electoral voting register, the findings may not be representative. This is because people under 18 years old, people who had recently moved in or out of the area, and people in temporary accommodation would be missed.

Differences between people asked to take part in a study and those who actually respond should also be considered before generalising findings to the target population. Non-participants can differ in important respects from participants. For example, if a study involves interviewing people at home during the day, those employed outside the home may be less likely to participate. The health and lifestyle of employed and non-employed people differs in many ways, so the findings could be misleading.

The proportion of individuals who drop out of a study after it has begun should be kept to a minimum. This is particularly important in cohort studies which follow people over long periods of time. People who drop out of studies often share common features. It is important to monitor the characteristics of the remaining participants to ensure that they are still representative of the target population.

Valid and reliable results

Any methods used must produce valid data. Measurements that provide information on what the study set out to measure, in other words data, are 'valid' if it measures what it is supposed to be measuring. If studying the effect of blood pressure on development of CVD, valid blood pressure measurements would be made using an appropriate blood pressure monitor. A survey to study the effect of alcohol consumption on the development of coronary heart disease could introduce problems with validity if it relied on the participants recalling the quantity of alcohol they consumed. Participants may not recall correctly because they were intoxicated, or they may underestimate because they are reluctant to admit their true consumption.

The method used to collect results must be reliable. A reliable method produces measurements that are repeatable and reproducible. The method will give similar results when repeated by one person using the same equipment and procedure under the same conditions over a short timescale. It will also give similar results when used at different times, or by different people. If measuring blood pressure, the same type of equipment and same procedure should be used each time the measurement is made. Any variables that could affect the measurement should be controlled or taken into account. A method using questionnaires, for example to conduct a survey on lifestyle factors, should use the same questions for each participant.

The disease diagnosis must be clearly defined, to ensure that different doctors record and measure symptoms in the same way. The development of coronary heart disease or onset of Alzheimer's, for example, must be measured and recorded using standard methods that are the same for all participants in the study.

ACTIVITY

In **Student Activity 1.13** you evaluate the design of studies used to determine health risk factors.

Sample size

A sample must be large enough to produce results that could not have occurred by chance. In cohort studies of a rare disease, only a small proportion of the population will develop the disease. In case-control studies, only a few people may have been exposed to the factors under investigation, or, in the case of rare diseases, the number of cases may be low to start with. With larger samples more accurate estimates for the wider population can be calculated. For a condition that affects 5% of the population each year, a cohort of 1000 people would need to be followed for 10 years in order to collect information on 50 people with the disease. Similarly, in case-control studies, sufficient participants need to be recruited in order to detect any effects due to rare exposures.

Controlling variables

The potential effect of all variables that could be correlated with the disease should be considered when designing the study. For example, in a study of blood pressure and development of CVD, a group of people with low blood pressure is compared with a group with higher blood pressure. The data shows that the group with lower blood pressure has less CVD. However, if the average age of this group is less than the high pressure group, the difference in CVD development may be due to the age difference and not blood pressure. Age is a factor known to be associated with CVD. Matching case and control groups on variables known to correlate with the disease being studied will ensure that only the factor under investigation is influencing the outcome.

Q 1.13 The Framingham cohort is primarily white. Explain whether it would be valid to extrapolate these results to the general population of the USA.

> **CHECKPOINT**
> **1.5** Produce a checklist of the features of a well-designed health risk study that ensure valid and reliable data are collected.

Risk factors for CVD

Your chances of having coronary heart disease or a stroke are increased by several inter-related risk factors, the majority of which are common to both conditions. These include:

- high blood pressure
- obesity
- blood cholesterol and other dietary factors
- smoking
- inactivity
- genetic inheritance.

Some of these you can control, while others you cannot.

Figure 1.26 Some of the potential risk factors for developing coronary heart disease are easy to identify, but may be difficult to control.

Age and gender make a difference

Q 1.14 Look at Table 1.1. What happens to your risk of developing cardiovascular disease as you get older?

Q 1.15 Does this mean that, at your age, you need not worry?

Age/years	Male		Female	
	Heart attack	Stroke	Heart attack	Stroke
16–44	1	1	1	1
45–64	19	8	8	3
65–74	47	17	31	12
75 and over	72	26	37	14

Table 1.1 Rates per 1000 population reporting longstanding diseases of the circulatory system by sex and age, 2011, Great Britain. *Source: Office for National Statistics General Household Survey.*

Q 1.16 Look at Table 1.2. Do these data suggest that males and females face the same risk of cardiovascular disease? Support your answer with calculated risk values.

	Male population (thousands)	Male deaths	Female population (thousands)	Females deaths
Under 35	13 959	193	13 620	98
35–44	4427	905	4513	297
45–54	4261	3054	4341	1011
55–64	3631	7025	3744	2308
65–74	2589	12 835	2820	6410
75+	1938	41 866	2916	53 932
Totals	30 805	65 878	31 954	64 056

Table 1.2 Population and mortality data from cardiovascular disease (coronary heart disease and stroke combined) for the United Kingdom in 2010. *Sources: Office for National Statistics Population Estimates for UK, mid-2010; British Heart Foundation/Department of Public Health University of Oxford CHD Statistics 2012 edition.*

Q 1.17 Comment on the reliability of the data presented in Tables 1.1 and 1.2.

Q 1.18 Many people now think that a woman's reproductive hormones offer her protection from coronary heart disease until they decline during the menopause in middle age when her monthly periods cease. Do these data support this view? Explain whether it is valid to draw this conclusion from these data.

⚙ ACTIVITY

In **Student Activity 1.14** you compare data for coronary heart disease and stroke and look at trends over a ten-year period.

The risk of cardiovascular disease is higher for men than women in the UK. In England in 2010 the incidence of heart attacks among men was 154 per 100 000, whereas the rate for women was only 34 per 100 000 (*Source: BHF 2012*). These figures are about a half of those recorded in 2002. In both sexes, the prevalence of cardiovascular disease (the proportion of cases in the population) increases with age. This may be due to the effects of ageing on the arteries; they tend to become less elastic and may be more easily damaged. With increasing age the risks associated with other factors may increase, causing a rise in the number of cases of disease.

Q 1.19 Suggest what might have caused the fall in heart attacks described above.

High blood pressure

Elevated blood pressure, known as **hypertension**, is considered to be one of the most common factors in the development of cardiovascular disease. High blood pressure increases the likelihood of atherosclerosis occurring.

Blood pressure is a measure of the hydrostatic force of the blood against the walls of a blood vessel. You should remember that blood pressure is higher in arteries and capillaries than in veins (Figure 1.28). The pressure in an artery is highest during the phase of the cardiac cycle when the ventricles have contracted and forced blood into the arteries. This is the **systolic pressure**. Pressure is at its lowest in the artery when the ventricles are relaxed. This is the **diastolic pressure**.

Measuring blood pressure

A **sphygmomanometer** is the traditional device used to measure blood pressure. It consists of an inflatable cuff that is wrapped around the upper arm, and a manometer, or gauge, that measures pressure (Figure 1.27). When the cuff is inflated the blood flow through the artery in the upper arm is stopped. As the pressure in the cuff is released the blood starts to flow through the artery. This flow of blood can be heard using a stethoscope positioned on the artery below the cuff. A pressure reading is taken when the blood first starts to spurt through the artery that has been closed. This is the *systolic* pressure. A second reading is taken when the pressure falls to the point where no sound can be heard and it equals the lowest pressure in the artery. This is the *diastolic* pressure.

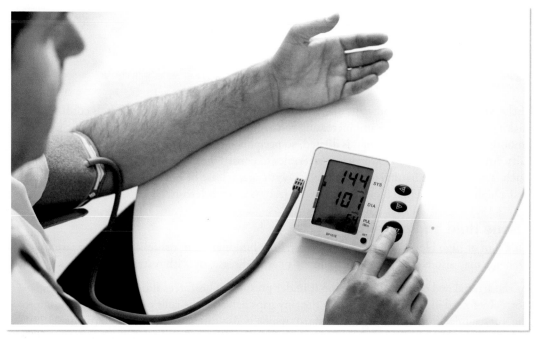

Figure 1.27 Nowadays blood pressure monitors can give digital readouts.

The SI units (International System of Units) for pressure are kilopascals, but in medical practice it is still traditional to use millimetres of mercury, mmHg. The numbers refer to the number of millimetres the pressure will raise a column of mercury.

Blood pressure is reported as two numbers, one 'over' the other, for example $\frac{140}{85}$. This means a systolic pressure of 140 mmHg and a diastolic pressure of 85 mmHg. For an average healthy person you would expect a systolic pressure of between 100 and 140 mmHg and a diastolic pressure of between 60 and 90 mmHg.

systolic pressure, the maximum blood → 140   diastolic pressure, the blood
pressure when the heart contracts $\overline{85}$ ← pressure when the heart is relaxed

Peter's blood pressure was an incredible $\frac{240}{140}$

> **☼ ACTIVITY**
>
> In **Student Activity 1.15** you use a sphygmomanometer, a blood pressure monitor, or the accompanying simulation to measure blood pressure.

What determines your blood pressure?

Contact between blood and the walls of the blood vessels causes friction and this impedes the flow of blood. This is called peripheral resistance. The arterioles and capillaries offer a greater total surface area than the arteries, resisting flow more, slowing the blood down and causing the blood pressure to fall. Notice in Figure 1.28 that the greatest drop in pressure occurs in the arterioles. The fluctuations in pressure in the arteries are caused by contraction and relaxation of the heart. As blood is expelled from the heart, pressure is higher. During diastole, elastic recoil of the blood vessels maintains the pressure and keeps the blood flowing.

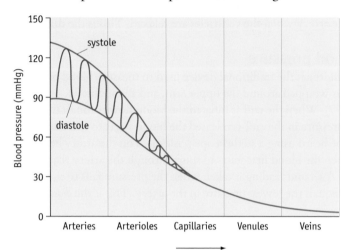

Figure 1.28 Blood pressure in the circulatory system. As peripheral resistance increases with greater total surface area, the flow of blood slows causing pressure to fall.

> **⚙ ACTIVITY**
> Draw a **concept map** for blood pressure to bring together all the ideas covered. A pro-forma is available in **Student Activity 1.16** if you do not want to start from scratch.

If the smooth muscles in the walls of an artery or an arteriole contract, the vessels constrict, making the lumen narrower and increasing resistance. In turn, your blood pressure is raised. If the smooth muscles relax, the lumen is dilated, so peripheral resistance is reduced and blood pressure falls. Any factor that causes arteries or arterioles to constrict can lead to elevated blood pressure. Such factors include natural loss of elasticity with age, release of hormones such as adrenaline, and a high-salt diet. High blood pressure can lead to atherosclerosis.

Tissue fluid and oedema

One sign of high blood pressure is **oedema** – fluid building up in tissues and causing swelling. Oedema may also be associated with kidney or liver disease, or with restricted body movement.

At the arterial end of a capillary, blood is under pressure. This forces fluid and small molecules normally found in plasma out through the tiny gaps between the cells of the capillary wall into the intercellular space, forming **tissue fluid**, which is also called interstitial fluid (Figure 1.29). Blood cells and larger plasma proteins stay inside the capillary; their larger size prevents them passing through the gaps in the capillary wall. The tissue fluid drains into a network of lymph capillaries which returns the fluid to the blood via a lymph vessel which empties into the vena cava.

If blood pressure rises above normal, more fluid may be forced out of the capillaries. In such circumstances, fluid accumulates within the tissues causing oedema.

Q 1.20 During left-side heart failure (the most frequent type) there is an increase in pressure in the pulmonary vein and left atrium. This is because blood continues to flow out of the right side of the heart to the lungs and return to the heart due to the action of breathing muscles whereas the left atrium and ventricle no longer pump blood out of the heart efficiently. Where in the body will blood pressure rise and oedema form?

A

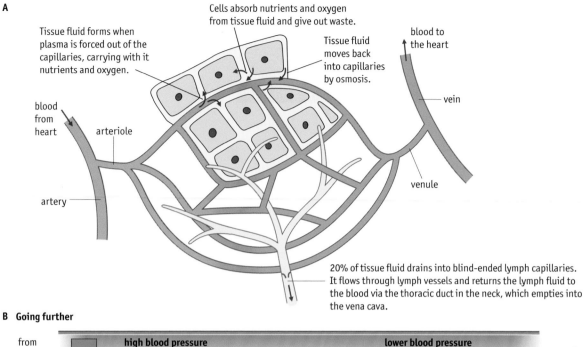

Tissue fluid forms when plasma is forced out of the capillaries, carrying with it nutrients and oxygen.

Cells absorb nutrients and oxygen from tissue fluid and give out waste.

Tissue fluid moves back into capillaries by osmosis.

blood to the heart

vein

blood from heart

arteriole

artery

venule

20% of tissue fluid drains into blind-ended lymph capillaries. It flows through lymph vessels and returns the lymph fluid to the blood via the thoracic duct in the neck, which empties into the vena cava.

B Going further

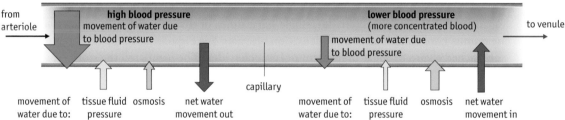

from arteriole

high blood pressure
movement of water due to blood pressure

lower blood pressure
(more concentrated blood)
movement of water due to blood pressure

to venule

capillary

movement of water due to: tissue fluid pressure   osmosis   net water movement out

movement of water due to: tissue fluid pressure   osmosis   net water movement in

Figure 1.29 A Production of tissue fluid in a capillary bed. **B** Notice how the net movement in is less than the movement out. The excess fluid formed is drained away through the lymphatic system. For details on osmosis see page 70.

Dietary factors

Our choices of food, in particular the type and quantity of high-energy food that we eat, can either increase or decrease our risk of developing certain diseases, including cardiovascular diseases.

Energy units – avoiding confusion

Most packet foods these days detail the energy content per 100 g or other appropriate quantity. Figure 1.30 shows the energy content for a bar of chocolate. Why are two different units of energy quoted? Which should we use?

Traditionally, energy was measured in **calories**; one calorie is the quantity of heat energy required to raise the temperature of 1 cm³ of water by 1 °C. Food labels normally display units of 1000 calories, called **kilocalories** (also called **Calories** with a capital C).

The SI unit for energy is the joule (J), and 4.18 joules = 1 calorie. The **kilojoule** (1 kJ = 1000 joules) is used extensively in stating the energy contents of foods. In the popular press the Calorie is still used as the basic unit of energy, particularly with reference to weight control. Hence most food labels in the UK continue to quote both Calories and kilojoules (Figure 1.30).

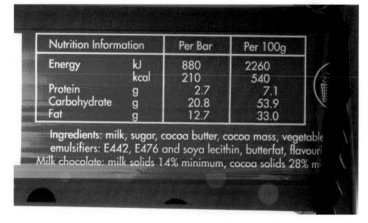

Nutrition Information		Per Bar	Per 100g
Energy	kJ	880	2260
	kcal	210	540
Protein	g	2.7	7.1
Carbohydrate	g	20.8	53.9
Fat	g	12.7	33.0

Ingredients: milk, sugar, cocoa butter, cocoa mass, vegetable emulsifiers: E442, E476 and soya lecithin, butterfat, flavouri Milk chocolate: milk solids 14% minimum, cocoa solids 28% mi

Figure 1.30 Nutritional information on a chocolate wrapper. How much energy does this chocolate contain? Notice that the label displays energy values in kilojoules and kilocalories.

Q 1.21   A newborn baby requires around 2000 kJ per day. Express this **a** in calories **b** in Calories.

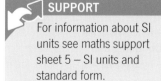

SUPPORT

For information about SI units see maths support sheet 5 – SI units and standard form.

Where do we get energy from in our diet?

Carbohydrates, **lipids** (often called fats and oils) and **proteins** are constituents of our food that contain energy. Alcohol can also provide energy. The relative energy content of these nutrients is shown in Table 1.3.

Nutrient	Energy available per gram/kJ
Carbohydrates	16
Lipids	37
Proteins	17
Alcohol	29

Table 1.3 Energy content of nutrients.

Carbohydrates

The term carbohydrate was first used in the nineteenth century and means 'hydrated carbon'. If you look at each carbon in a carbohydrate molecule (see Figure 1.33) you should be able to work out why, bearing in mind that hydration means adding water.

Most people are familiar with sugar and starch being classified as carbohydrates, but the term covers a large group of compounds with the general formula $Cx(H_2O)n$.

Sugars are either **monosaccharides**, single sugar units, or **disaccharides**, in which two single sugar units have combined in a condensation reaction. See Figures 1.31 and 1.32. Long straight or branched chains of sugar units form **polysaccharides**. The names tell the story – mono means one, di – two, and poly – many.

> **ACTIVITY**
>
> Complete the interactive tutorial in **Student Activity 1.17** to help you understand carbohydrate structure.

monosaccharides can be joined by condensation reactions to form

disaccharides

and **polysaccharides** containing three or more sugar units.

Figure 1.31 A simplified diagram to show how simple sugar units (monomers) can be joined to form more complex carbohydrates (polymers).

KEY BIOLOGICAL PRINCIPLE: LARGE BIOLOGICAL MOLECULES ARE OFTEN BUILT FROM SIMPLE SUBUNITS LINKED IN CONDENSATION REACTIONS

All organisms rely on the same basic building blocks as a result of our shared evolutionary origins. Hydrogen, carbon, oxygen and nitrogen account for more than 99% of the atoms found in living organisms. Relatively simple molecules join together in different ways to produce many of the large important biological molecules.

Polymers, such as polysaccharides (Figure 1.31), proteins and nucleic acids, are made by linking identical or similar subunits, called monomers, to form straight or branched chains. Lipids are another group of biological molecules also constructed by joining smaller molecules together, though they are not polymers since they are not chains of monomers. Large biological molecules have structures that are well suited to their functions.

In each case, the small molecules join together in a condensation reaction, so called because a water molecule is released as the two molecules combine in the reaction (see Figure 1.32). **Condensation** reactions are common in the formation of complex molecules. Addition of water in a **hydrolysis** reaction splits the molecule (Figure 1.32 and 1.36).

In Topic 1 we are looking at the structure and function of some carbohydrates and lipids, returning in later topics to see how these molecules have many other roles. In Topic 2 the structure and function of nucleic acids and proteins will be examined in detail.

Figure 1.32 A simplified diagram to show the condensation and hydrolysis reactions involved in the formation or splitting of a disaccharide. In this example it is the reaction between two glucose molecules. Full details can be found in Figure 1.34.

Monosaccharides

Monosaccharides are single sugar units with the general formula $(CH_2O)n$, where n is the number of carbon atoms in the molecule. Monosaccharides have between three and seven carbon atoms, but the most common number is six. For example, the monosaccharides **glucose**, **galactose** and **fructose** all contain six carbon atoms and are known as **hexose** sugars (Figure 1.33).

Q 1.22 What are the ratios of carbon, hydrogen and oxygen in monosaccharides?

A hexose sugar molecule has a ring structure formed by five carbons and an oxygen atom; the sixth carbon projects above or below the ring. The carbon atoms in the molecule are numbered, starting with 1 on the extreme right of the molecule. The side branches project above or below the ring, and their position determines the type of sugar molecule and its properties.

Monosaccharides provide a rapid source of energy. They are readily absorbed and require little or, in the case of glucose, no change before being used in cellular respiration. Glucose and fructose are found naturally in fruit, vegetables and honey; they are both used extensively in cakes, biscuits and other prepared foods.

Q 1.23 At first glance galactose and glucose look similar. Compare their molecular structures shown in Figure 1.33 and describe how they differ.

Glucose is important as the main sugar used by all cells in respiration. Starch and glycogen are polymers made up of glucose subunits joined together. When starch or glycogen is digested, glucose is produced. This can be absorbed and transported in the bloodstream to cells. This is known as α glucose, there is another form you will study in Topic 4.

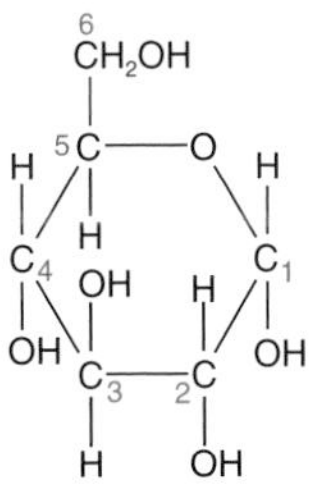

This can also be shown more simply by omitting the Cs in the ring:

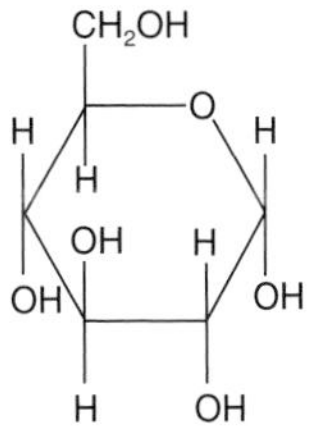

SUPPORT
For information about chemical reactions, bonds and carbohydrates see the biochemistry support on the website.

Galactose occurs in our diet mainly as part of the disaccharide sugar lactose, which is found in milk. Notice that the –OH groups on carbon 1 and carbon 4 lie on the opposite side of the ring compared with their position in glucose.

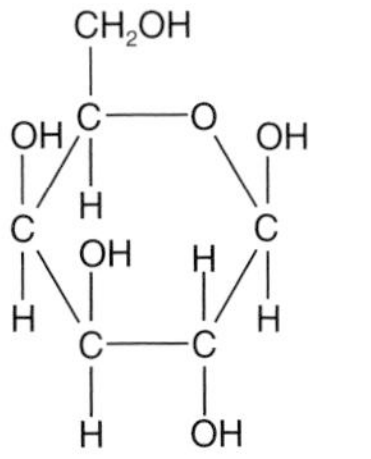

Fructose is a sugar which occurs naturally in fruit, honey and some vegetables. Its sweetness attracts animals to eat the fruits and so help with seed dispersal.

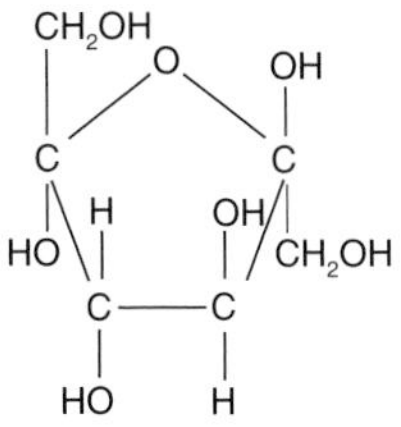

Figure 1.33 Glucose, galactose and fructose are examples of monosaccharides. They are all hexose sugars.

Disaccharides

Two single sugar units can join together and form a disaccharide (double sugar) in a **condensation** reaction releasing a water molecule as the two sugar molecules combine in the reaction. The bond that forms between the two sugar units is known as a **glycosidic bond** or link. Figure 1.34 shows the formation of the disaccharide maltose by a condensation reaction between two glucose molecules.

The bond in maltose is known as a 1,4 glycosidic bond because it forms between carbon 1 on one molecule and carbon 4 on the other. Remember that the carbons are numbered anticlockwise from the oxygen.

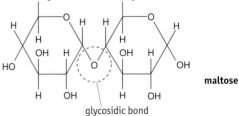

maltose

glycosidic bond

Figure 1.34 Two glucose molecules may join in a condensation reaction to form the disaccharide maltose. A water molecule is released during the reaction.

Common disaccharides found in food are **sucrose**, **maltose** and **lactose**. Their structures are shown in Figure 1.35.

Sucrose
Sucrose, formed from glucose and fructose, is the usual form in which sugar is transported around the plant.

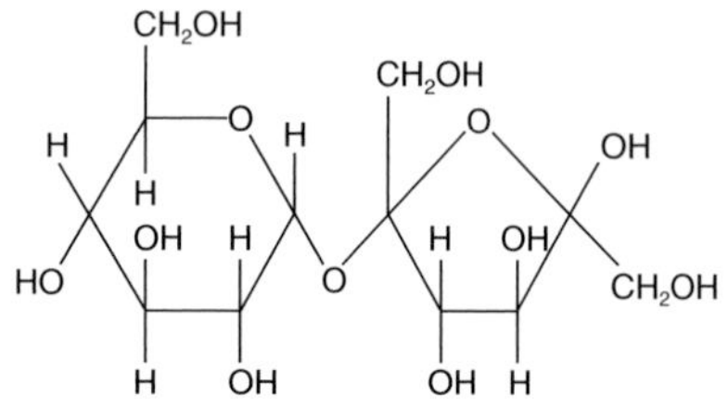

glucose   fructose

Maltose
Maltose, formed from two glucose molecules, is the disaccharide produced when amylase breaks down starch. It is found in germinating seeds such as barley as they break down their starch stores to use for food.

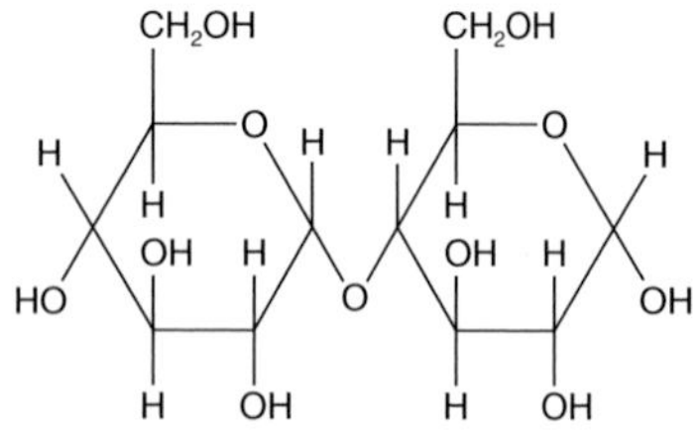

glucose   glucose

Lactose
Galactose and glucose make up lactose, the sugar found in milk.

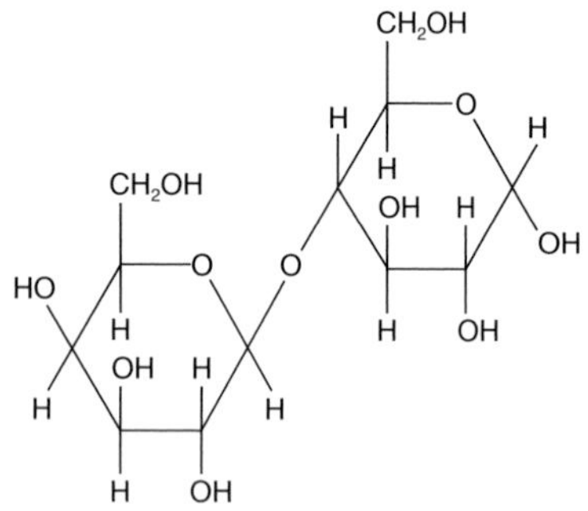

glucose   galactose

Figure 1.35 Disaccharides formed by joining two monosaccharide units.

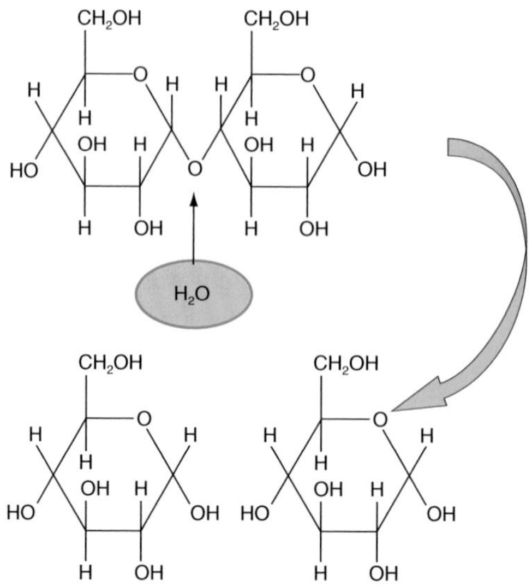

H_2O

Figure 1.36 The glycosidic bond between the two glucose molecules in maltose can be split by hydrolysis. In this reaction water is added.

Q 1.24 Identify the glycosidic bond in each molecule shown in Figure 1.35.

(a) sucrose

(b) maltose

(c) lactose

The white or brown crystalline sugar we use in cooking, and the sugar in golden syrup or molasses, is sucrose. It is extracted from sugar cane or sugar beet.

The glycosidic link between two sugar units in a disaccharide can be split by **hydrolysis**. This is the reverse of condensation: water is added to the bond and the molecule splits into two (Figure 1.36). Hydrolysis of carbohydrates takes place when carbohydrates are digested in the gut and when carbohydrate stores in a cell are broken down to release sugars.

Q 1.25 Using the molecule in Figure 1.35, sketch what the monomers that make up lactose would look like after hydrolysis.

If monosaccharides are eaten they are rapidly absorbed into the blood causing a sharp rise in blood sugar. Polysaccharides and disaccharides (complex carbohydrates) have to be digested into monosaccharides before being absorbed, which takes some time, so the monosaccharides are released more slowly. Eating complex carbohydrates does not cause the swings in blood sugar levels we see after eating monosaccharides.

Lactose is the sugar present in milk. Many adults are intolerant of lactose and drinking milk will produce unpleasant digestive problems for these people. Asian and Afro-Caribbean people have a particularly high rate of lactose intolerance. One solution is to hydrolyse the lactose in milk, which converts the *disaccharide* lactose into the *monosaccharides* glucose and galactose. Industrially this is carried out using the enzyme lactase. Lactase can be immobilised in a gel, and milk is poured in a continuous stream through a column containing beads of the immobilised enzyme (Figure 1.37).

> **ACTIVITY**
>
> In **Student Activity 1.18** you immobilise lactase and use it to hydrolyse lactose.

Figure 1.37 Lactose free milk is produced industrially by pouring milk through a column containing the enzyme lactase immobilised in gel beads. The lactase catalyses the hydrolysis of the lactose. Syrup which is used in the food industry is produced in the same way from whey waste from cheese-making.

DID YOU KNOW?

Why do we have such a sweet tooth?

We have taste receptors on the tongue for five main tastes – sweet, sour, bitter, salty and umami (the taste associated with monosodium glutamate or MSG). It is likely that the sweet-taste receptors enable animals to identify food that is easily digestible, whereas bitter-taste receptors provide a warning to avoid potential toxins. Humans, along with many other primates (apes and monkeys), have many more sweet-taste receptors than most other animals. Our sweet-taste receptors help us to identify when fruit is ready to eat.

Polysaccharides

Polysaccharides are polymers made up from simple sugar monomers joined by glycosidic bonds into long chains, as shown in Figure 1.38. Each sugar monomer is joined to the chain in a condensation reaction with a water molecule released during the reaction.

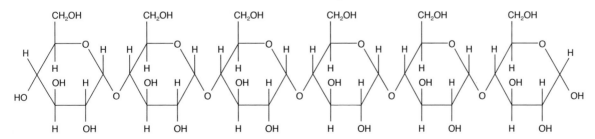

Figure 1.38 Glycosidic bonds join the glucose molecules that make up this polysaccharide.

Q 1.26 How many water molecules will have been released in the formation of the molecule in Figure 1.38?

There are three main types of polysaccharide found in food: **starch** and **cellulose** in plants, and **glycogen** in animals. Although all three are polymers of glucose molecules, they are sparingly soluble (they do not dissolve easily) and do not taste sweet.

Starch and glycogen act as energy storage molecules within cells. These polysaccharides are suitable for storage because they are compact molecules with low solubility in water. This means that they do not affect the concentration of water in the cytoplasm and so do not affect movement of water into or out of the cell by osmosis. See Topic 2 for details of osmosis.

Starch, the storage carbohydrate found in plants, is made up of a mixture of two molecules, **amylose** and **amylopectin**.

● Amylose is composed of a straight chain of between 200 and 5000 glucose molecules with 1,4 glycosidic bonds between adjacent glucose molecules. The position of the bonds causes the chain to coil into a spiral shape.

● Amylopectin is also a polymer of glucose but it has side branches. A 1,6 glycosidic link holds each side branch onto the main chain.

Figure 1.39 attempts to show these complex 3D structures. Starch grains in most plant species are composed of about 70–80% amylopectin and 20–30% amylose. The compact spiral structure of starch and its insoluble nature make it an excellent storage molecule. It does not diffuse across cell membranes and has very little osmotic effect within the cell.

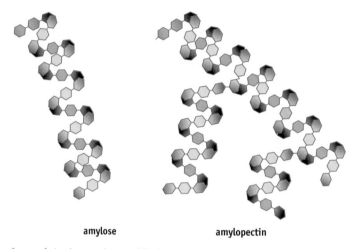

amylose   amylopectin

Figure 1.39 The two forms of starch – amylose and the branched chain amylopectin. The chains of glucose molecules coil to form a spiral. This is held in place by hydrogen bonds that form between the hydroxyl (OH) groups which project into the centre of the spiral.

Starch is a major source of energy in our diet and is common in many foods (Figure 1.40). It occurs naturally in fruit, vegetables and cereals, often in large amounts. The sticky gel formed when starch is mixed with water makes it a good thickening agent and it is also added to many food products as a replacement for fat.

Figure 1.40 Foods high in starch.

Glycogen is used by bacteria, fungi and animals instead of starch as an energy store. It is another polymer composed of glucose molecules. Its numerous side branches (Figure 1.41) mean that it can be rapidly hydrolysed, giving easy access to stored energy. In humans glycogen is stored in the liver and muscles.

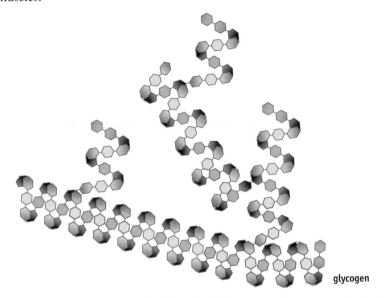

glycogen

Figure 1.41 Glycogen, the storage carbohydrate found in animal cells, has a branched structure similar to amylopectin.

CHECKPOINT

1.6 Produce a table of comparison, differences and similarities, between monosaccharides, disaccharides and polysaccharides.

Cellulose in the diet is known as **dietary fibre**, and it is also referred to as a non-starch polysaccharide. Up to 10 000 glucose molecules are joined to form a straight chain with no branches (the glucose molecules have a slightly different structure to those found in starch). The structure and function of cellulose are considered in Topic 4; it is not required at this stage.

Indigestible in the human gut, cellulose has an important function in the movement of material through the digestive tract. Dietary fibre is thought to be important in the prevention of 'Western diseases' such as coronary heart disease, diabetes and bowel cancer.

Lipids

Lipid is the general term for fats and oils. In food, lipids enhance flavour and palatability, making it feel smoother and creamier (Figure 1.42). They supply over twice the energy of carbohydrates with 37 kJ of energy per gram of food. This can be an advantage if large amounts of energy need to be consumed in a small mass of food. It also means a large amount of energy can be stored in a small mass, for example in seeds.

Figure 1.42 Which is more popular — with or without fat?

Lipids are organic molecules found in every type of cell. They are insoluble in water but soluble in organic solvents such as ethanol. Most of the lipids that we eat are **triglycerides** which are used as energy stores in plants and animals. Triglycerides are made up of three fatty acids and one glycerol molecule linked by condensation reactions (Figure 1.43). The bond that forms between each fatty acid and the glycerol is known as an **ester bond**. Three ester bonds are formed in a triglyceride. Each is formed in a condensation reaction with the release of a water molecule.

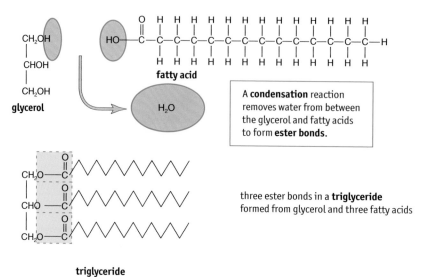

Figure 1.43 The formation of a triglyceride, a common type of lipid. The fatty acid chains in the triglyceride are simplified, each point in the zig-zag shows the position of a carbon atom.

Saturated fats

If the fatty acid chains in a lipid contain the maximum number of hydrogen atoms they are said to be **saturated**. In a saturated fatty acid the hydrocarbon chain is long and straight (Figure 1.44).

There are no carbon to carbon double bonds in the saturated fatty acid chain and no more hydrogens can be added to it. Animal fats from meat and dairy products are major sources of saturated fats.

Straight, saturated hydrocarbon chains can pack together closely. The strong intermolecular bonds between triglycerides made up of saturated fatty acids result in fats that are solid at room temperature.

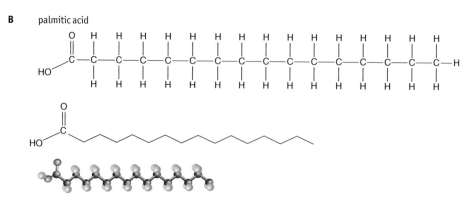

Figure 1.44 A A saturated hydrocarbon chain. **B** Palmitic acid, a saturated fatty acid with a straight hydrocarbon chain containing 16 carbons in total. It can be drawn in several ways with or without the carbons and hydrogens in the chain, or as a 3D model.

Unsaturated fats

Monounsaturated fats have one double bond between two of the carbon atoms in each fatty acid chain (Figure 1.45). **Polyunsaturated** fats have a larger number of double bonds. A double bond causes a kink in the hydrocarbon chain. These kinks prevent the unsaturated hydrocarbon chains packing closely together. Increasing the distance between the molecules weakens the intermolecular forces between the unsaturated triglycerides resulting in oils that are liquid at room temperature. Olive oil is particularly high in monounsaturated fats. Most other vegetable oils, nuts and fish are good sources of polyunsaturated fats.

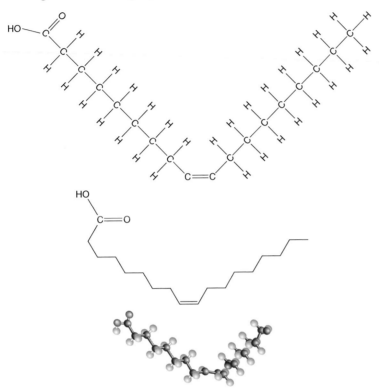

Figure 1.45 Oleic acid, an unsaturated fatty acid, has a double bond which causes a kink in the hydrocarbon chain. It can be drawn in several ways.

Unsaturated fats can be made more solid at room temperature by adding hydrogen to the double bonds making them saturated. These hydrogenated-, or trans-, fats are sometimes produced by the food industry and used in processed foods. Trans-fats do occur naturally at very low levels in meat and dairy products.

Other types of lipid

Cholesterol (Figure 1.46) is a short lipid molecule which is essential for good health. It is a vital component of cell membranes with roles in their organisation and functioning. The steroid sex hormones (such as progesterone and testosterone) and some growth hormones are made

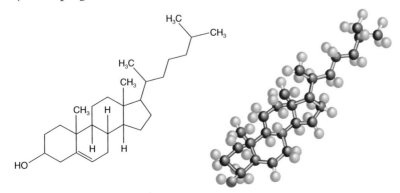

Figure 1.46 Alternative ways of showing the structure of cholesterol.

ACTIVITY

Student Activity 1.19
Complete this ICT-based tutorial to understand lipid structure.

from cholesterol. Bile salts, involved in lipid digestion and assimilation, are also formed from cholesterol. Cholesterol is made in the liver from saturated fats and also obtained in our diet. It is found associated with saturated fats in foods such as eggs, meat and dairy products. However, there are concerns that a high blood cholesterol level can be bad for us.

Phospholipids are similar to triglycerides but one of the fatty acids is replaced by a negatively charged phosphate group. In Topic 2 you will study the detailed structure of phospholipids and their role in forming cell membranes.

Fats provide more than just energy

As well as supplying energy in the diet fats also provide a source of **essential fatty acids**, that is, fatty acids that the body needs but cannot synthesise. Fats must therefore be present in a balanced diet to avoid deficiency symptoms. For example, a deficiency of linoleic acid (an essential fatty acid) can result in scaly skin, hair loss and slow wound healing. In addition, the fat-soluble vitamins (A, D, E and K) can only be absorbed if our diet includes food containing fat.

The energy balance

Look on food labels and you will often see recommended daily amounts for nutrients along with daily energy requirements for men and women. But how much energy is right for each of us, and what happens if we do not get it right?

Getting it right

The UK Department of Health publishes dietary guidelines for most nutrients. They used to give recommended daily amounts but in 1991 these were largely replaced with dietary reference values (DRVs). DRVs are estimates of requirements and are not recommendations or goals for individuals. DRVs include:

● an estimated average requirement (EAR)

● a lower reference nutrient intake (LRNI)

● a higher reference nutrient intake (HRNI).

These effectively provide a range of values within which a healthy balanced diet should fall (Figure 1.47). Upper and lower limits have not been set for carbohydrates and fats. Instead, estimated average requirements for energy are suggested plus the average percentage that should come from the different energy components of a diet. Tables 1.4 and 1.5 give the recommendations.

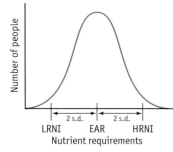

Figure 1.47 For a group of people who receive adequate nutrients, the range of intakes will vary around the EAR.

SUPPORT

The LRNI and HRNI lie two standard deviations from the EAR. For more information on standard deviation see maths support 10 – standard deviation.

WEBLINK

The British Nutrition Foundation website provides detailed information on nutrient requirements and recommendations.

Q 1.27 Look at Table 1.4. Compare the energy requirements of males and females, commenting on changes with age.

Estimated average requirements (EARs) for energy per day				
Age	Males		Females	
	(kJ/day)	(kcalories/day)	(kJ/day)	(kcalories/day)
15 years	11 800	2820	10 000	2390
16 years	12 400	2964	10 100	2414
17 years	12 900	3083	10 300	2462
18 years	13 200	3155	10 300	2462
19–24 years	11 600	2772	9100	2175
25–34 years	11 500	2749	9100	2175
35–44 years	11 000	2629	8800	2103
45–54 years	10 800	2581	8800	2103
55–64 years	10 800	2581	8700	2079
65–74 years	9800	2342	8000	1912
75+ years	9600	2294	7700	1840

Table 1.4 Estimated average requirements for energy in kJ per day. The recommendation for the general adult population is 2000 kcal for women and 2500 kcal for men. *Source: Scientific Advisory Committee on Nutrition (SCAN), 2011.*

% of daily total food energy intake excluding alcohol			
Fat		Carbohydrates	
Saturated	Unsaturated	Starch	Sugars
11	24	39	11

Table 1.5 Dietary guidelines for percentage of daily energy that should come from carbohydrate and fat. Alcohol should provide no more than 5% of energy in the diet.

Getting it wrong

We have to be aware that we need both carbohydrates and fats in our diet for good health, but that there are consequences if we get it wrong by consuming too much energy or if the percentage supplied by the various components differs greatly from the guidelines.

You need a constant supply of energy to maintain your essential body processes, such as the pumping of the heart, breathing and maintaining a constant body temperature. These processes go on all the time, even when you are completely at rest. The energy needed for these essential processes is called the **basal metabolic rate** (BMR) and varies between individuals. BMR is higher in:

- males
- heavier people
- younger people
- more active people.

The estimated average requirements for energy shown in Table 1.4 were calculated by multiplying basal metabolic rate by a physical activity level factor of 1.4, which reflects current average levels of physical activity. This gives estimates suitable for people who do little physical activity at work or in leisure time.

Q 1.28 The requirements in Table 1.4 can be altered to give estimates for very active people by using a physical level factor of 1.9. Use this value to work out what the EAR of a very active 17 year old male and female.

Adult females usually require about 8400 kJ a day and adult males usually need about 10 500 kJ a day, but an athlete may require double this quantity or even more (Figure 1.49).

If you eat fewer kilojoules per day than you use you have a negative energy balance and energy stored in the body will be used to meet the demand. A regular shortfall in energy intake will result in weight loss. If you routinely eat more energy than you use, you have a positive energy balance. The additional energy will be stored and you will put on weight (Figure 1.48).

ACTIVITY

Student Activity 1.20 uses dietary analysis software to help you to work out your own energy budget and determine whether you are getting the right amount of energy from the best sources.

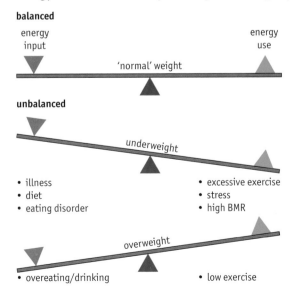

Figure 1.48 The balance between energy input and energy output determines whether the body maintains, gains or loses weight.

Figure 1.49 The winner of the Tour de France 2014, Vincenzo Nibali. During the tour the cyclists use about 33 000 kJ a day.

In England it is estimated that over 65% of men (16 years and over) and over 55% of women (16 years and over) are either overweight or obese. Approximately 25% of the adult population are obese (2010 – 26%) and **obesity** in men has tripled since the mid-1980s 6–8%. The 2010 Health Survey of England reported that about 30% of 2–15-year-olds were overweight, including 13.7% who were obese. The increasing prevalence of obesity among both children and adults has been called an epidemic in Western Europe and the USA.

Defining 'overweight' and 'obese'

Body mass index (BMI) is a conventionally used method of classifying body weight relative to a person's height. To calculate BMI, body mass (in kg) is divided by height (in metres) squared:

$$\text{BMI} = \frac{\text{body mass/kg}}{\text{height}^2/\text{m}^2}$$

For example, the BMI of a person with a body mass of 65 kg and height of 1.72 m is 22.0. This figure can then be used to identify the category of body weight to which that person belongs, as shown in Table 1.6. BMI does not have an exact correlation with fat levels in the body. BMI may not be accurate for athletes, children, people over 60, or those with long-term health conditions.

> **⚙ ACTIVITY**
>
> Work out BMI and waist-to-hip ratio using **Student Activity 1.21**.

BMI	Classification of body weight
<18.5	Underweight
18.5–24.9	Normal
25.0–29.9	Overweight
30.0–40.0	Obese
>40.0	Severely obese

Table 1.6 The use of BMI to classify body weight.

Q 1.29 **(a)** Calculate the body mass index of a person with a body mass of 85 kg and height of 1.68 m. How would you describe the body weight of this person?

(b) Rajesh is 191 cm tall. His BMI is 30. How much does he weigh in kg?

There is evidence that **waist-to-hip ratio** is a better measure of obesity than BMI and shows a highly significant association with risk of heart attack. A large-scale case-control study, the INTERHEART Study, looked at 27 098 people in 52 countries; 12 461 of these people had already had a heart attack and they were matched with control individuals of the same sex and similar age who had no history of cardiovascular disease. The participants were asked about their health, economic status, lifestyle and family history of coronary heart disease. Their weight, height, waist and hip circumference were measured to allow BMI and waist-to-hip ratios to be calculated.

The results found that BMIs in men and women who had previously had a heart attack were only slightly higher than the BMIs of the control group. However, their waist-to-hip ratios were much higher than those of the control group. There is a continuous positive correlation between waist-to-hip ratio and heart attack (Figure 1.50). Waist-to-hip ratio gives a better indication of who is at risk of a heart attack, even in people with BMIs of less than 20.

Waist-to-hip ratio is calculated by dividing waist circumference by hip circumference. The waist is measured unclothed at the narrowest point between the rib margin and the top of the hip bone. The hip circumference is measured in light clothing at the widest point around the buttocks. A non-stretchable tape measure is used attached to a spring scale with a tension of 750 g. Ideally men should not have a waist-to-hip ratio over 0.90 and women's should not be greater than 0.85.

Q 1.30 A 45-year-old man has a waist measurement of 91 cm and a hip measurement of 115 cm. Calculate his waist-to-hip ratio and comment on his risk of heart disease.

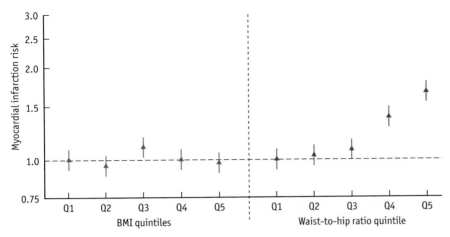

Figure 1.50 The relationship between waist-to-hip ratio and heart attack.

A poor diet, particularly one high in sugar and fat, and a sedentary lifestyle are the major contributing factors to the development of obesity. There is evidence in the UK that fat consumption has actually declined since 1990, but greater inactivity means that obesity and associated conditions are on the increase.

A high-fat diet will not necessarily result in weight gain if combined with a high level of physical activity. This was clearly illustrated by Ranulph Fiennes and Mike Stroud during their 1993 expedition to cross the Antarctic on foot – they found 23 000 kJ a day inadequate to meet their energy needs (Figure 1.51).

Consequences of obesity

Obesity increases your risk of coronary heart disease and stroke, even without other risk factors being present. The more excess fat you carry, especially around your middle, the greater the risk to your heart. Obesity can also greatly increase your risk of type II diabetes. Type II diabetes is also referred to as non-insulin-dependent diabetes or late-onset diabetes. It, in turn, increases your risk of coronary heart disease and stroke.

Figure 1.51 Over 23 000 kJ a day and Ranulph Fiennes still lost weight! Even with a padded jacket he looks thin.

DID YOU KNOW?

What is type II diabetes?

There are about 3 million people with diabetes in the UK and probably at least another 850 000 people with diabetes who do not know they have the condition. Type I occurs when the body is unable to make insulin.

In type II diabetes the body either does not produce sufficient insulin or the body fails to respond to the insulin that is produced. You probably know that **insulin** is the hormone that helps regulate **blood glucose levels**. After a meal the level of blood glucose rises. In response to this change, the **pancreas** produces insulin and secretes it into the bloodstream. The insulin causes cells to absorb glucose, and therefore the blood glucose level returns to normal. Continually high blood glucose levels due to frequent consumption of sugar-rich foods can reduce the sensitivity of cells to insulin, resulting in type II diabetes. It may take years to develop and may not even be diagnosed. It is thought that up to a million people in the UK could be unaware that they have type II diabetes. Type II is the more common of the two main types of diabetes and accounts for between 85 and 95% of all people with diabetes.

Obesity can also raise your blood pressure and elevate your blood lipid levels, two classic risk factors for cardiovascular disease. Much media attention, particularly in advertising, is focused on saturated fats and cholesterol.

Why is cholesterol such a problem?

There is a considerable amount of evidence to show that the higher your blood cholesterol level, the greater your risk of coronary heart disease (Figure 1.52A). The British Regional Heart Study reported similar results, with blood cholesterol having a log-linear relationship with CHD risk (Figure 1.52B).

Q 1.31 Using the data in Figure 1.52A, comment on the relationship between serum cholesterol levels and the risk of death from coronary heart disease.

SUPPORT

To find out more about using logs look at maths support sheet 2 Presenting data – graphs on the website.

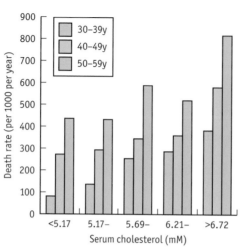
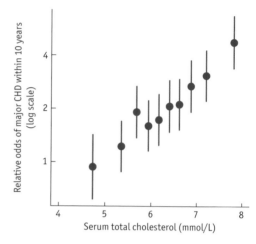

Figure 1.52 A The effect of blood cholesterol levels on the incidence of coronary heart disease. Data from the Framingham Study. **B** Relative odds of major CHD within 10 years by usual serum total cholesterol. Estimates are presented for each of ten equal groups and are adjusted for age, smoking status, diastolic blood pressure and serum total cholesterol.

The World Health Organisation has estimated that in high-income countries, 50% of CVD events can be attributed to blood cholesterol levels over 3.8 mmol/l.

The INTERHEART case control study estimated that in Western Europe, 45% of heart attacks are due to an abnormal blood cholesterol level.

However, as you may realise, it is not quite as simple as that! Like all lipids, cholesterol is not soluble in water. In order to be transported in the bloodstream, insoluble cholesterol is combined with proteins to form soluble **lipoproteins**.

There are two major transport lipoproteins.

● **Low-density lipoproteins (LDLs)** – triglycerides from fats in our diet combine with cholesterol and protein to form LDLs which transport the cholesterol to body cells. LDLs circulate in the bloodstream and bind to receptor sites on cell membranes before being taken up by the cells where the cholesterol is involved in the synthesis and maintenance of cell membranes. Excess LDLs overload these membrane receptors, resulting in high blood cholesterol levels. This LDL cholesterol may be deposited in the artery walls forming atheromas.

● **High-density lipoproteins (HDLs)** – HDLs have a higher percentage of protein and less cholesterol compared with LDLs, hence their higher density. High-density lipoproteins are made when triglycerides from fats combine with cholesterol and protein. HDLs transport cholesterol from the body tissues to the liver where it is broken down. This lowers blood cholesterol levels and helps remove the fatty plaques of atherosclerosis.

LDLs are associated with the formation of atherosclerotic plaques whereas HDLs reduce blood cholesterol deposition. Therefore, it is desirable to maintain a higher level of HDL (the so-called 'good cholesterol' or 'protective cholesterol') and a lower level of LDL (the so-called 'bad cholesterol'); a higher HDL:LDL ratio in the blood.

Saturated fats versus unsaturated fats

Studies have shown that saturated fat in the diet increases LDL and HDL cholesterol, however, the increase in LDL cholesterol is greater. Studies have also reported that although replacing saturated fat with polyunsaturated fat decreases both LDL and HDL levels, a greater reduction in LDLs means that the HDL:LDL ratio is increased with a protective effect. It is recommended that eating a low-fat diet that particularly avoids saturated fats will help reduce total blood cholesterol, and especially LDL cholesterol, which constitutes the major component of the cholesterol risk for CVD. Saturated fats may also reduce the activity of LDL receptors so the LDLs are not removed from the blood, thus further increasing the blood cholesterol levels and CVD risk

Q 1.32 Up until the menopause, women generally have higher HDL:LDL ratios than men. What consequences would you expect this to have for the incidence of coronary heart disease in women compared with men?

Q 1.33 A person stops eating butter on their toast and starts using a 'lighter' butter instead that contains 25% vegetable oil. What effect will this have on their blood LDL levels? Explain your answer.

Q 1.34 It been suggested that HDLs may reduce platelet aggregation. Explain why this might reduce the risk of a heart attack occurring.

Measuring blood LDL level is difficult and expensive, but it is relatively easy to measure the total cholesterol level. You can even test your own cholesterol at home now, using a home test kit (Figure 1.53). Testing for total cholesterol may be part of a cardiovascular assessment for someone suspected of being at risk of CVD.

Conflicting evidence

Many studies provide evidence for a positive correlation between fat consumption and CHD mortality rates (Figure 1.52). However, conflicting evidence has also been reported. For example, in France CHD is low despite high intake of cholesterol and saturated fat. The mean total blood cholesterol levels in France and the UK are similar but the mortality rate due to CHD is much lower in France.

An analysis of 72 cohort and randomised control studies into fatty acid intake and CHD published in 2014 concluded that there was no significant association between saturated fat and coronary disease, or evidence that polyunsaturated fats have a protective effect. However, the authors noted limitations in the studies and the need for further study. It did conclude that there was a significant risk associated with trans-fats, which are polyunsaturated oils hydrogenated to make them more solid for use in processed food industry. The Scientific Advisory Committee in Nutrition recommend that trans-fats should not exceed 2% of food energy. Since 2008, UK supermarkets and fast food chains have stopped using trans-fats. Food manufactured outside of the UK could still contain trans-fats.

Q 1.35 Several hypotheses have been put forward to explain the difference in CHD in France and the UK. Suggest what might cause the difference.

Smoking

Smoking cigarettes is one of the major risk factors for the development of cardiovascular disease. A 50-year cohort study of British doctors found that CHD mortality was 60% higher in smokers compared with non-smokers. The constituents in smoke affect the circulatory system in the following ways.

- Carbon monoxide in the smoke binds to the haemoglobin (the oxygen-carrying protein in red blood cells) instead of oxygen. This reduces the supply of oxygen to cells. This will result in an increased heart rate as the body reacts to provide enough oxygen for the cells.

- Nicotine in smoke stimulates the production of the hormone adrenaline. This hormone causes an increase in heart rate and also causes arteries and arterioles to constrict, both of which raise blood pressure.

> **⚙ ACTIVITY**
> In **Student Activity 1.22** you look at some evidence for a causal relationship between blood cholesterol levels and cardiovascular disease and some conflicting evidence.

Figure 1.53 A home cholesterol kit.

● The numerous chemicals that are found in smoke can cause damage to the lining of the arteries, triggering atherosclerosis.

● Smoking has also been linked with a reduction in HDL cholesterol level.

Q 1.36 Suggest why doctors might prescribe nicotine patches if nicotine causes an increase in blood pressure?

Inactivity

The British Heart Foundation considers physical inactivity to be one of the most common risk factors for heart disease. They estimate that only three or four in every 10 men and two or three in every 10 women do sufficient exercise to give some protection against heart disease. It has been shown that being active can halve the risk of developing coronary heart disease.

Moderate exercise such as walking, cycling or swimming, helps prevent high blood pressure and can help to lower it. Exercise not only helps maintain a healthy weight, it also seems to raise HDL cholesterol without affecting LDL cholesterol levels. It also reduces the chance of developing type II diabetes and helps in controlling the condition.

A person who is physically active is much more likely to survive a heart attack or stroke compared with someone who has been inactive. The Department of Health recommends adults undertake two and a half hours of moderate to vigorous physical activity each week and aim to do some exercise each day.

My dad had a heart attack – will I?

If one or other of your parents suffers or suffered from cardiovascular disease, you are more likely to develop it yourself. There may be inherited predisposition for the disease.

Heredity and risk

Some diseases, such as some simple genetic disorders, have a single risk factor. These diseases are determined by the inheritance of a defective allele, and the risk of suffering from the conditions follows the Mendelian rules of inheritance that you have already met at GCSE.

For example, consider two people who are carriers of sickle cell anaemia. This is a recessive genetic condition in which an abnormal form of haemoglobin is produced. This abnormal haemoglobin is less soluble and at low oxygen concentrations it crystallises, causing the red blood cells to become distorted. The sickle- or crescent-shaped cells are less efficient at carrying oxygen and can block small blood vessels.

In Figure 1.54, two carriers decide to try to have a child. The chances of this child, or any others they have, inheriting the defective form of the gene from both parents is 1 in 4.

	Mother		Father	
Parent genotype	Aa		Aa	
Parent phenotype	normal red blood cells		normal red blood cells	
Gametes	A	a	A	a

	A	a
A	AA	Aa
a	Aa	(aa)

Any child with this genotype will have sickle cell anaemia.

Figure 1.54 If both parents carry one defective allele there is a 1 in 4 chance of any child inheriting two copies of the defective allele and suffering from sickle cell anaemia.

With some genetically inherited conditions the risks are very clear cut. However, inheritance is often much more complex. Even in some conditions that are controlled by a single gene, it is now known that different mutations of that gene determine how severe the person's condition is. This is true for cystic fibrosis, which you will study in detail in Topic 2.

Some diseases result from several genes interacting. In other diseases, genes have been identified that do not *cause* the condition directly but that still increase the individual's chance of developing it. There is no clear-cut relationship between having these genes and having the condition; rather the genes *increase* the individual's susceptibility to the disease.

Genes and CHD

There are some single gene disorders that increase the likelihood of early development of coronary heart disease. For example, in familial hypercholesterolaemia (FH), mutations in the LDLR gene cause the LDL receptors, involved in removal of LDL from the blood, not to form or to have a shape that makes them less efficient. This results in high blood LDL levels with early onset of CHD. It is thought that about 1 in 500 people carry a mutation in this gene and it may account for 5–10% of the coronary artery disease in people below 55 years of age. However, even with this gene mutation the development of CHD will still be dependent on other environmental factors, such as lifestyle and diet.

Apolipoprotein gene cluster

The inheritance of cardiovascular disease is rarely a simple case of a single faulty gene for the condition being passed from one generation to the next. There are several genes that can affect your likelihood of developing cardiovascular disease. The apolipoprotein gene cluster has been identified as associated with coronary heart disease and other conditions such as Alzheimer's disease. This group of genes has been extensively researched, and it appears that some alleles are linked to higher risk whereas others may reduce the risk.

Apolipoproteins are the protein component of lipoproteins. They are mostly formed in the liver and intestines and have important roles in stabilising the structure of the lipoproteins and recognising receptors involved in lipoprotein uptake on the plasma membrane of most cells in the body. There are several types of apolipoproteins, including the three described below.

● Apolipoprotein A (APOA) – the major protein in HDL, which helps in the removal of cholesterol to the liver for excretion. Mutations in the apoA gene are associated with low HDL levels and reduced removal of cholesterol from the blood, leading to increased risk of coronary heart disease.

● Apolipoprotein B (APOB) – the main protein in LDL, the molecule that transfers cholesterol from the blood to cells. Mutations of the apoB gene result in higher levels of LDL in the blood and a higher susceptibility to CVD.

● Apolipoprotein E (APOE) – a major component of HDLs and very low-density lipoproteins (VLDLs), which are also involved in removal of excess cholesterol from the blood to the liver. The apoE gene has three common alleles, producing three forms of the protein, E2, E3 and E4. APOE4 slows removal of cholesterol from the blood and therefore having the E4 allele may increase the risk of coronary heart disease.

Q 1.37 How many phenotypes of APOE are possible?

Q 1.38 One mutation of APOA, known as Milano (because it is common in people from Milan in Italy), results in very low levels of HDL, but each HDL is very efficient at removing cholesterol from the blood. How will this affect the risk of developing CVD for those who are carriers of the mutation?

APO gene mutations and CHD risk

There are numerous mutations in the APO genes and the effects of these mutations are modified in different environments and when in different combinations, making it impossible to estimate the effect of a single gene or single environmental risk factor. In the future, testing for levels of apolipoproteins in a person's blood may be a better indicator of a person's risk of CHD than cholesterol levels themselves.

ACTIVITY
You can read about the role of genes in the sudden death of athletes in **Student Activity 1.23**.

CHD is multifactorial

The chance that a person will suffer from cancer or cardiovascular disease, for example, is rarely the consequence of genetic inheritance alone. Many such diseases are **multifactorial**, with heredity, the physical environment, the social environment and lifestyle behaviour choices *all* contributing to the risk. The combination of risk factors experienced by the individual determines their risk of developing the disease.

Families do not pass on just their genes. You may also acquire your parents' lifestyle and its associated risk factors, such as smoking, lack of exercise and poor diet. Genetic tests are available for alleles associated with single gene disorders that increase the likelihood of early development of coronary heart disease (Figure 1.55). For the majority of people genetic testing for alleles associated with increased risk of CHD is of little value due to the interaction of risk factors.

Other risk factors

The most important factors that increase the risk of cardiovascular disease are smoking, having high blood pressure, having a high level of blood cholesterol and lack of physical activity. As we have seen, obesity is also a major risk factor. In addition, there are other things you should think about when deciding what to do if you want to lower your CVD risk.

The role of antioxidants

During reactions in the body, unstable radicals result when an atom has an unpaired electron. For example, the superoxide radical, which is oxygen with an unpaired electron; this is often represented as $O_2\cdot$. Radicals (sometimes known as free radicals) are highly reactive and can damage many cell components including enzymes and genetic material. This type of cellular damage has been implicated in the development of some types of cancer, heart disease and premature ageing. Some vitamins, including vitamin C, beta-carotene and vitamin E, can protect against radical damage. They provide hydrogen atoms that stabilise the radical by pairing up with its unpaired electron. The MONICA study (mentioned on page 23) found that high levels of antioxidants seemed to protect against heart disease. Partly because of their function as a good source of antioxidants, the current Department of Health recommendation is to include at least five portions of fruit or vegetables per day in our diet. Analysis of 2008 Health Survey of England data showed that people eating seven or more a day had a 33% reduced risk of death due to CVD or cancer compared with people who ate less than one.

Wine and some fruit juices contain chemicals that have antioxidant properties and also help stop platelets sticking together.

Q 1.39 How might the antioxidants in wine help reduce the incidence of cardiovascular disease?

Salt

You need some salt in your diet for the healthy functioning of your body but too much can have adverse effects. The Food Standards Agency recommends a salt intake of no more than 6 g per day for an adult, but the UK average intake is double that figure. Approximately 75% of the salt eaten comes from processed food. An average-sized bowl of breakfast cereal can contain as much as 1 g of salt; a standard-sized bag of crisps about 0.5 g. Often the value for sodium rather than salt is given; the value for salt is about 2½ times greater. A high-salt diet causes the kidneys to retain water. Higher fluid levels in the blood result in elevated blood pressure with the associated cardiovascular disease risks.

Stress

How you respond to stress in your life is very important. There is evidence that coronary heart disease is sometimes linked to poor stress management. In stressful situations the release of adrenaline causes arteries and arterioles to constrict, resulting in raised blood pressure. Stress can also lead to overeating, a poor diet and higher alcohol consumption, which are all potential contributors to CVD.

Figure 1.55 Genetic testing is becoming more common. Some companies now provide such testing so that individuals can see if they are at high or low risk of developing certain diseases.

ACTIVITY

Student Activity 1.24 lets you determine if your diet contains enough antioxidant vitamins.

ACTIVITY

Student Activity 1.25 lets you investigate the quantity of the antioxidant vitamin C in fruit juice.

ACTIVITY

Student Activity 1.26 is a teacher-led demonstration that lets you take part in an investigation of some factors that affect blood pressure and heart rate.

Alcohol

Heavy drinkers are at far greater risk of heart disease and a number of other diseases. Heavy drinking raises blood pressure, contributes to obesity and can cause irregular heartbeat. However, there has also been much research and debate concerning *potential* protective effects of *moderate* drinking.

If you have a glass of wine, the alcohol it contains (1 unit or 8 g) is very quickly absorbed, 20% through the wall of the stomach and the remainder through the walls of the small intestine. Excess alcohol consumption can result in direct tissue damage, including damage to the liver, brain and heart. Such damage contributes to an increased risk of cardiovascular disease. The liver has many functions but two of its main ones are processing carbohydrates, fats and proteins, and detoxification, including the removal and destruction of alcohol. High levels of alcohol can damage liver cells. This impairs the ability of the liver to remove glucose and lipids from the blood. In the liver alcohol is converted into ethanal, a three-carbon carbohydrate. Most of the ethanal is used in respiration but some may end up in very low-density lipoproteins (VLDLs), increasing the risk of plaque deposition.

Given these harmful consequences of drinking alcohol, it seems remarkable to claim that moderate drinking may actually offer some degree of protection against cardiovascular disease. However, studies have shown a small protective effect of alcohol, in particular wine, compared with abstinence. This is thought to be because moderate alcohol consumption is correlated with higher HDL cholesterol levels (the 'good' cholesterol).

If you do drink, moderation is the key! The UK recommended limits to avoid health problems are 2–3 units per day for women, and 3–4 units per day for men, with no binge drinking. There is one unit of alcohol in half a pint of average strength beer, a small glass of table wine or a measure of spirits (Figure 1.56). One unit of alcohol is approximately the amount that an adult eliminates from the body in one hour.

Coffee

Epidemological studies have positively and negatively correlated drinking coffee and health risk. Some studies suggest increased CVD risk whereas others suggest moderate consumption may have health benefits.

Figure 1.56 Many larger glasses of wine available in pubs, restaurants and wine bars contain 2–3 units of alcohol, the recommended daily limit for women to avoid health problems.

> **⚙ ACTIVITY**
> In **Student Activity 1.27** you can find out whether caffeine increases heart rate and blood pressure.

> **⚙ ACTIVITY**
> In **Student Activity 1.28** you can 'Test your healthy heart IQ'.

> **✓ CHECKPOINT**
> **1.7** Produce a concept map or table which shows the risk factors for CVD and their effects.

<u>1.4</u> Reducing the risks of cardiovascular disease

The risk of cardiovascular disease can be reduced in a range of different ways, including:

● stopping smoking

● maintaining resting blood pressure below 140/85 mmHg

● maintaining low blood cholesterol level

● maintaining a normal BMI/low waist-to-hip ratio

● taking regular physical exercise

● moderate or no use of alcohol.

If people in the UK did not smoke, the British Heart Foundation estimates that 10 000 fewer men and women of working age would die from heart attacks each year. After stopping smoking, your risk of coronary heart disease is almost halved after only one year.

Controlling blood pressure

If a person is diagnosed with high blood pressure, changes in diet and lifestyle would be recommended. Medications are also available to reduce high blood pressure. These would normally be prescribed for people with sustained systolic pressure of ≥160 mmHg or sustained diastolic pressure of ≥100 mmHg. People with sustained blood pressure over 140/90 mmHg and who show evidence of CVD may also be treated.

Three main types of drugs are used to treat high blood pressure.

ACE inhibitors

ACE inhibitors (angiotensin converting enzyme inhibitors) are effective antihypertensive drugs which reduce the synthesis of angiotensin II. This hormone causes vasoconstriction of blood vessels to help control blood pressure. The ACE inhibitors prevent the hormone being produced from the inactive angiotensin I, therefore reducing vasoconstriction and lowering blood pressure.

Some people experience side effects when taking ACE inhibitors. These include a dry cough, dizziness due to rapid lowering of blood pressure, abnormal heart rhythms and a reduction in the function of the kidney. However, for anyone with kidney disease the drug may reduce the risk of kidney failure. Patients intolerant to ACE inhibitors will be prescribed an alternative drug that blocks the angiotensin II receptors.

Calcium channel blockers

Calcium channel blockers are antihypertensive drugs that block the calcium channels in the muscle cells in the lining of arteries. For the muscle to contract, calcium must pass through these channels into the muscle cells. Failure of calcium to enter the cell prevents contraction of the muscle, the blood vessels do not constrict, and this lowers blood pressure. In Topic 7 you will look at muscle contraction and the role of calcium in more detail.

There are some side effects with these drugs, such as headaches, dizziness, swollen ankles due to a build-up of fluid in the legs, abnormal heart rhythms, flushing red in the face and constipation. In people with heart failure, taking some types of calcium channel blockers can make symptoms worse or be fatal, so they may not be prescribed to people who have had a heart attack.

Diuretics

Diuretics increase the volume of urine produced by the kidneys and therefore rid the body of excess fluids and salt. This leads to a decrease in blood plasma volume and cardiac output (volume of blood expelled from the heart in a minute), which lowers blood pressure.

A few people taking diuretics may have some side effects, such as dizziness, nausea or muscle cramps. If you are taking a diuretic it is important not to have too much salt in food as this counteracts the diuretic effect.

Most people take two of these drugs to control blood pressure, usually an ACE inhibitor in combination with one of the others. In 2006 the British Hypertension Society and the National Institute for Clinical Excellence (NICE) published guidelines on the appropriate selection of these drugs to use, with the drugs selected according to what is effective and carries the lowest risk of side effects.

Reducing blood cholesterol levels

There is evidence from both the UK and the USA that untargeted cholesterol screening of the general population combined with dietary advice has little effect on lowering blood cholesterol levels. A further problem is associated with 'labelling' people; when told that they have high blood pressure, many people react by signing off sick!

Q 1.40   Why do you think untargeted screening and dietary advice are not very effective at lowering blood cholesterol?

Some people need to reduce blood cholesterol. One way to achieve this is through a low-fat diet. The media constantly remind us of the need to do this.

Table 1.7 shows the extent of cholesterol-lowering that is obtained from following a low-fat diet in high-risk patients, namely people who have already experienced a heart attack, compared with the general population.

	Blood cholesterol reduction	
	mmol/l	%
general population	0.22	3
high-risk patients	0.65	9

Table 1.7 The effect of lipid-lowering diets in reducing blood cholesterol levels.

Q 1.41   Can you suggest why the effect of dietary change appears to have been more successful in reducing blood cholesterol in patients who had experienced a heart attack than in the general population?

Cholesterol-lowering drugs

Individuals diagnosed with high cholesterol who have CVD or have a 20% chance of developing CVD in the next 10 years may also be prescribed cholesterol-lowering drugs. The most commonly prescribed are statins. Statins work by inhibiting an enzyme involved in the production of LDL cholesterol by the liver. Side effects of statins are tiredness, disturbed sleep, feeling sick, vomiting, diarrhoea, headache and muscle weakness.

Several large studies have investigated the effects of these drugs and researchers have collaborated to analyse the results of 14 randomised trials. The results, drawing on 90 000 participants, were published in 2005, and showed that statins quickly reduce the risk of heart disease and stroke. Studies have shown that there is a linear relationship between absolute reductions in LDL cholesterol and reductions in the incidence of major vascular events such as heart attack and stroke. In patients who have CVD, treatment with statins can reduce the risk of heart attack by up to 33%. In people who do not have CVD but have elevated cholesterol levels, statins lower total and LDL cholesterol by more than 20% and the risk of CVD by a similar amount.

Some studies have raised concerns that the use of statins may increase the risk of death by other non-vascular causes, and also increase the risk of developing particular cancers, such as gastrointestinal and respiratory cancers. To determine if this was the case, researchers in the UK and Australia analysed the results of 14 large randomised trials for statins. They found no evidence that lowering LDL cholesterol by 1 mmol/l with five years of statin treatment increased the risk of death by other non-vascular causes or of developing particular cancers. However, they did highlight the need for extended study of patients using statins over long periods to identify any adverse effects that might occur with long-term use.

There is an ongoing debate (2014) about whether the use of statins should be extended to those at low risk of CHD, as advocated by the University of Oxford Clinical Trials Service and Health Economics Research Centre study that looked at the results of randomised trials involving 175 000 participants for an average of five years.

A diet to reduce the risk of cardiovascular disease

A diet to offer protection against cardiovascular disease would include the following key features (Figure 1.57).

- Energy balanced.
- Reduced saturated fat.
- More polyunsaturated fats.
- Reduced cholesterol.
- Reduced salt.
- More non-starch polysaccharides, such as pectins and guar gum.

These polysaccharides, known as soluble fibre, have been found to lower blood cholesterol. They are found in fruit, vegetables, beans, pulses and some grains (e.g. oats). They are only partially digested, forming a gel that traps the cholesterol and prevents its absorption.

Figure 1.57 A diet to reduce the risk of developing cardiovascular disease.

● Includes oily fish.

Fish such as mackerel, sardines, anchovies, salmon and trout contain omega-3 fatty acids, a group of polyunsaturated fatty acids with their first double bond between the third and fourth carbon atoms. These fatty acids are essential for cell functioning and have been linked to a reduction in heart disease and joint inflammation. The evidence for the importance of omega-3 fatty acids is seen in the Inuit in Greenland and the inhabitants of certain Japanese islands. They regularly eat oily fish and have very low rates of coronary heart disease.

● More fruit and vegetables.

Fruits and vegetables contain antioxidants and often non-starch polysaccharides.

Studies suggest that including foods containing about 2 g/day of plant sterols can reduce LDL cholesterol levels by about 9%. These are naturally produced substances in plants, similar to cholesterol. They compete with cholesterol during its absorption in the intestine. Unfortunately, we would have to eat vast quantities of foods such as vegetable oils and grains in order to reduce our cholesterol levels through this competition. Products have therefore been developed that can incorporate sterols into everyday foodstuffs such as yogurts. However, NICE consider there is currently (2014) insufficient evidence for them to recommend them as they are unlikely to give additional benefit for the majority of people over a low-fat diet. There are no studies to show an impact on the occurrence of CHD.

REVIEW
In **Student Extension**
1.3 you can find out more about functional food trials.

Figure 1.58 Some yoghurts and margarines include sterols to help lower cholesterol. These products are known as functional foods.

Anticoagulant and platelet inhibitory drug treatment

If someone has had a heart attack or stroke, or is identified as being at high risk of one, in addition to the lifestyle changes and drug treatments to reduce blood pressure and blood cholesterol, they may be given drugs to prevent formation of a blood clot in an artery.

The tendency for platelet aggregation and clotting is reduced by platelet inhibitory drugs and anticoagulant drugs. Aspirin reduces the stickiness of platelets and the likelihood of clot formation. But some people are allergic to aspirin, and for others it is not effective or is only partially effective. In these cases an alternative platelet inhibitory drug, clopidogrel, may be used. A combined treatment involving a daily dose of aspirin and clopidogrel can have a dramatic effect. However, it has been shown that there is a risk of bleeding in the gastrointestinal tract with aspirin, and in trials there have been high rates of serious bleeding when aspirin is used in combination with clopidogrel. This risk of bleeding may outweigh the benefits. If a person has only a low risk of a vascular event (less than 1% per year risk of a heart attack or stroke), the risk of bleeding outweighs the benefits. The risks and benefits need to be considered for each individual patient, though guidelines suggest use when the 10-year risk of a heart attack or stroke is greater than 20%.

Warfarin is an anticoagulant drug. It interferes with the production of vitamin K and therefore it affects the synthesis of clotting factors. It can be taken orally for extended periods of time to prevent clotting. The benefits may be greater than with aspirin for some patients but the risk of bleeding is higher than with aspirin.

Q 1.42 **(a)** Reeta has a high risk of heart attack and is discussing drug treatments with her doctor. What information would they have to consider before making a decision about the best drug to use?

(b) Regular low doses of aspirin reduce the risk of bowel cancer. Would it be sensible for everyone to take low doses of aspirin to reduce their risk of cancer? Give a reason for your answer.

Q 1.43 Look back at the section on blood clotting then explain how warfarin prevents blood clotting.

Think back to Mark and Peter. What is most surprising is that Mark was only 15 when he had his stroke. Mark had no obvious risk factors that would have alerted him to the possibility of having a stroke. He reports having taken exercise, eaten a reasonably healthy diet and having not smoked. If you read Mark's full story in Activity 1.1 at the start of the topic, you will find that he had a type of stroke in which a blood vessel supplying blood to the brain bursts.

An artery can burst due to an aneurysm, where blood builds up behind a section of artery that has narrowed as a result of atherosclerosis. However, with no risk factors and in one so young, the likelihood of atheroma deposits having built up to this extent in Mark's arteries seems unlikely. Although there was no history of stroke in Mark's family, he thinks that he inherited an allele causing him to have thin artery walls more prone to bursting.

Q 1.44 What additional information would you need from Mark to determine if his stroke was due to this type of inherited condition or due to atherosclerosis?

Mark remains healthy today (Summer 2014) and has not had a recurrence of the problem. Peter was lucky to be alive having had a blood pressure of 240/140 mmHg, two heart attacks and heart surgery (Figure 1.55). If you look back at his story you will recall that his father died aged 53 from a heart attack. This suggests that there may have been an inherited predisposition for the condition. Thankfully, Peter's active lifestyle will have helped him survive for many years after his surgery.

ACTIVITY
In **Student Activity 1.29** you can discuss how people use scientific information to reduce their risk of coronary heart disease.

EXTENSION
You can read about new treatments for coronary heart disease in **Student Extension 1.4**. 'New treatments for cardiovascular disease' provides a fine start. It even gives you the opportunity to observe surgery!

Haemorrhagic stroke

Blood vessels on the surface of the brain and those within the brain are susceptible to bursting resulting in a stroke. A haemorrhagic stroke occurs when a blood vessel supplying blood to the brain bursts. If the burst occurs within the brain it is known as an *intracerebral haemorrhage*, whereas bursting of a vessel on the surface causes what is known as a *subarachnoid haemorrhage*. Look at Figure 1.59 and work out why the different types of stroke were given these names. There is no need for you to remember the names, but it is worth being aware of them as doctors and medical scientists use this sort of language.

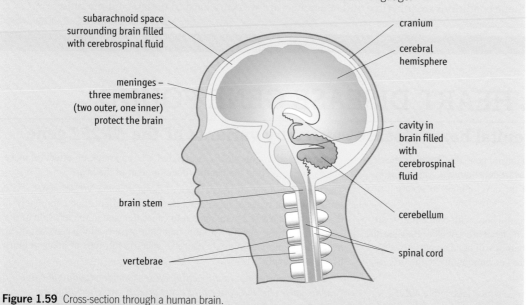

Figure 1.59 Cross-section through a human brain.

Figure 1.60 A vein taken from the patient's leg is used to bypass the sections of coronary artery that are narrowed. The photograph above shows the section of vein that has been grafted onto a patient's heart.

Q 1.45   What advice would you recommend Peter gives to his two daughters and son?

THINKING BIGGER

GENETIC DEFECTS OF THE HEART

We tend to think of heart disease as being a problem of older age due to atherosclerosis, largely unaware that some babies are born with heart disease. This is known as congenital heart disease; it refers to a heart defect or condition that is present at birth. There are many different types of congenital heart disease with some being minor and easily treated. whereas others are more serious. Some conditions are inherited and researchers are working hard to understand the causes.

8 April 2014

CONGENITAL HEART DISEASE GENE FOUND

Severe forms of congenital heart disease caused by variants of the *NR2F2* gene

Researchers have explored the role of a master gene that controls the functioning of other genes involved in heart development. Variations in this gene – *NR2F2* – are responsible for the development of severe forms of congenital heart disease.

Approximately one per cent of all babies are born with congenital heart disease, where the normal workings of the heart are affected. Because the damage to the heart is structural, most babies will need surgery to correct the problem. Although genetic causes are known to underlie the disease, these causes are not very well understood.

Scientists have previously shown that mice with a less active *NR2F2* gene had abnormal heart development. To see if the gene was involved in severe forms of human congenital heart disease, the team looked at DNA sequences of parents and affected children and found that variation on the *NR2F2* caused the structural damage that underlies these conditions.

The team found that these genetic variants were typically only present in the child and not the parents, revealing that congenital heart disease producing variants occur in the womb.

"What we see is that these rare variants in the NR2F2 gene interfere with the normal heart development and cause severe forms of congenital heart disease during human development," says Saeed Al Turki, first author from the Wellcome Trust Sanger Institute.

NR2F2 is a master regulator for other genes involved in the development of a healthy functioning heart

– once the activity of *NR2F2* is affected it has a knock-on effect on these other genes affecting the healthy development of the heart.

The team found that different types of damage in the *NR2F2* gene cause different types of heart defects. Genetic variants that completely deactivate the *NR2F2* gene tended to cause damage to the left side of the heart. In contrast, genetic variants that alter activity of the gene but do not deactivate it more commonly caused a specific sub-type of holes in the hearts of patients.

"With this knowledge, we are getting closer to understanding the full genetic causes behind congenital heart disease, which will provide better diagnoses and in turn provide better patient management," says Dr Matthew Hurles, senior author from the Wellcome Trust Sanger Institute.

Publication details
Rare variants in NR2F2 cause congenital heart defects in humans. Al Turki S, Manickaraj AK, Mercer CL, Gerety SS, Hitz MP, Lindsay S, D'Alessandro LC, Swaminathan GJ, Bentham J, Arndt AK, Low J, Breckpot J, Gewillig M, Thienpont B, Abdul-Khaliq H, Harnack C, Hoff K, Kramer HH, Schubert S, Siebert R, Toka O, Cosgrove C, Watkins H, Lucassen AM, O'Kelly IM, Salmon AP, Bu'lock FA, Granados-Riveron J, Setchfield K, Thornborough C, Brook JD, Mulder B, Klaassen S, Bhattacharya S, Devriendt K, Fitzpatrick DF, UK10K Consortium, Wilson DI, Mital S and Hurles ME *American journal of human genetics* 2014; **94**; 4; 574–85 *Press release published on the Wellcome Trust Sanger Institute website at http://www.sanger.ac.uk/ about/press/2014/140408.html*

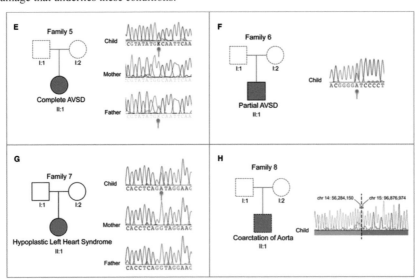

Family charts and sequencing results of NR2F2 variants in eight families affected by congenital heart disease (part of the diagram is shown above). Solid lines in pedigree charts indicate both whole-exome sequencing data and capillary sequencing data are available; dashed lines indicate samples with NR2F2 capillary sequencing data only.

Links across the course

1.1   1.2   1.3   1.4   YOU ARE HERE   2.1   2.2   2.3   2.4   2.5   2.6

Command words
Note that when the word critically is used in this context it does not mean that one should necessarily critise, it means that you should express your reasoned judgement.

START BY REVIEWING THE SOURCE

This article about the finding of a gene for congenital heart disease comes from the Wellcome Trust Sanger Institute website.

1. Read the article and comment on who you think the article might be aimed at.

2. Comment critically on the reliability of the article as a source of scientific information.

Biological vocabulary
As you read the article identify any unfamiliar words. Look these up to check you understand their meaning, you could look in the SNAB online glossary however if they are more specialised terms use the Internet to find a definition, making sure that the website you access is reliable, it is worth looking at a range of sources to check.

REVIEWING THE BIOLOGY

Having read the article, draw on your knowledge gained so far in the course and answer the following questions.

1. Explain in detail what the presence of the genetic variant in the child and not the parent tells you about how and where the variant may have arisen.

2. The article figure shows that most of the babies in these families had a congenital defect known as AVSD (Atrioventricular Septal Defect). These babies have a defect in their septum – the wall between the left and right sides of the heart. They have a hole through their septum between the atria and between the ventricles, with only a common atrioventricular valve between the atria and ventricles as shown in the diagram below. Using your knowledge of the function of the heart describe how these defects in the heart are likely to affect the circulation of blood. Think carefully about the pressure within the heart.

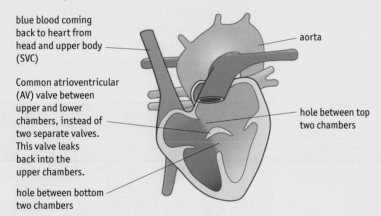

blue blood coming back to heart from head and upper body (SVC)

Common atrioventricular (AV) valve between upper and lower chambers, instead of two separate valves. This valve leaks back into the upper chambers.

hole between bottom two chambers

aorta

hole between top two chambers

3. A baby born with coarctation of aorta has narrowing of the aorta just beyond the branches that supply the head and arms. as occurred in family H. Suggest how the baby might be affected by this narrowing.

Once you have completed the remainder of the course come back and have a go at answering the following questions.

1. The figure caption refers to family charts, what would the left hand part of each diagram normally be called and describe what the circles and squares on this part of the diagram represent.

2. Suggest what might be meant by variant in the article?

3. Look at the figure and decide if you can work out what has happened to the DNA of each child in families one to four.

4. Explain the role of a master gene in control of development.

To find out more detail about congenital heart defects and how they can be treated visit the British Heart Foundation website.

GENES AND HEALTH

Why a topic called Genes and health?

It is now recognised that our genes have a major part to play in our health. It is estimated there are between 4000 and 6000 genetic disorders, which means that 1 in 25 children is affected and around 30 000 babies and children in the UK are diagnosed with a genetic disorder each year. Some disorders are minor and create little or no problem, for example, colour blindness and hairy ears are both inherited genetic conditions. Others have more serious consequences for the affected individuals. Albinism, cystic fibrosis, Huntington's disease, sickle cell disease and haemophilia are all serious conditions caused by faulty genes.

Thankfully, most serious genetic disorders are rare, but when one does occur it has a significant effect on the person's life and on the lives of their family members. Not only must affected individuals deal with the condition itself, they may also face difficult decisions about the possible inheritance of the condition by the next generation.

Cystic fibrosis dilemma

In this topic we follow a young couple, Claire and Nathan, who face a daunting dilemma – should they start a family when possibly their child could have cystic fibrosis (Figure 2.1)? Claire's sister Rachel has the condition, so Claire and Nathan think it might also be passed on to any child they have. To make the decision about whether to have a baby they need more information. How is cystic fibrosis inherited? What are the chances of any child they have inheriting the condition? If the child does inherit cystic fibrosis, how will it affect the child's life and theirs? Can genetic screening help? These are just some of the questions that need answers if Nathan and Claire are to make an informed choice.

Figure 2.1 Will this be the outcome for Claire and Nathan? Will they decide to have a baby? Will their baby have CF?

Cystic fibrosis (CF) is one of the most common genetic diseases, affecting approximately 9500 people in the UK and 100 000 people worldwide. One in 25 of the population carries the faulty CF allele. Every week, on average five babies are born with CF and two young people die from it, usually as a result of lung damage. In the 1960s the average life expectancy for a child with CF was just 5 years. By the year 2000 life expectancy had risen to 31 and today (2015) the CF Trust say the median predicted survival is 41 years old, so half the UK's cystic fibrosis population will live past 41. Research advances mean that newborns today are very likely to live even longer.

Claire is well aware of the outward symptoms of CF. She sees the problems her sister Rachel has with breathing, including a troublesome cough and repeated chest infections. Rachel also has to be very careful about her diet to ensure she can overcome problems with digestion and maintain her weight. Claire and Nathan want to know in more detail what causes these outward symptoms.

The symptoms of CF concern a sticky mucus layer that lines many of the tubes and ducts in the gas exchange, digestive and reproductive systems. The symptoms and the way CF is inherited have been known for a long time. The condition was first accurately described as cystic fibrosis of the pancreas by Dorothy Andersen, a New York pathologist, in 1938. She also developed one of the diagnostic tests which is still in use today. However, the gene responsible for the disease was not identified until 1989 by a group in Toronto led by Lap-Chee Tsui.

WEBLINK

To find out the life expectancy for someone with CF today visit the Cystic Fibrosis Trust website.

SUPPORT

To find out why the CF Trust use a median rather than an average or mean life expectancy see the Maths support sheet on averages on the website.

OVERVIEW OF THE BIOLOGICAL PRINCIPLES COVERED IN THIS TOPIC

In this topic you will study how changes in DNA can result in genetic disease using the example of cystic fibrosis. You will first look in detail at the symptoms of cystic fibrosis, extending your previous knowledge of the structure of the lungs and of gas exchange to see the importance of surface area-to-volume ratios in biology.

To understand the symptoms of cystic fibrosis you will study cell membrane structure and how substances move across membranes. To explain how faults arise in the cell surface membranes of a person with CF you will learn how genes code for proteins, how proteins are made and how their functions are dependent on their structure. You will find out how genes play their role in inheritance.

You will discover how genetic screening may be used to help people make decisions about the possible inheritance of genetic conditions. Throughout the topic you will consider the ethical issues raised by these technologies and you will learn how ethical arguments can be evaluated.

REVIEW

Are you ready to tackle Topic 2 *Genes and health*?

Complete the GCSE reviews and GCSE review tests before you start.

⚙ ACTIVITY

Student Activity 2.1 will give you an overview of cystic fibrosis, and let you find out more about Claire's family and the problems she and Nathan could face in the future.

Read the stories in **Student Activity 2.2** to see how cystic fibrosis has affected some people, and how they and their families cope.

2.1 The effects of CF on the lungs

The role of mucus in the lungs

The lungs allow rapid gas exchange between the atmosphere and the blood. Air is drawn into the lungs via the trachea due to low pressure in the lungs, created by an increase in the volume of the thorax as the ribs move up and diaphragm moves down. When the diaphragm muscles and those between the ribs relax, volume decreases, pressure rises and air is forced out through the trachea. The trachea divides into two bronchi which carry air to and from each lung. Within each lung there is a tree-like system of tubes ending in narrow tubes, bronchioles, attached to tiny balloon-like alveoli (Figures 2.2 and 2.5). The alveoli are the sites of gas exchange.

There is nothing unusual about having a layer of **mucus** in the tubes of the gas exchange system. Everyone normally has a thin coating of mucus in these tubes that is produced continuously from goblet cells in the walls of the airways (Figure 2.3). Any dust, debris or microorganisms that enter the airways become trapped in the mucus. This mucus is continually removed by the wave-like beating of **cilia** that cover the **epithelial** cells lining the tubes of the gas exchange system. However, people with CF, like Rachel, have mucus that contains less water than usual resulting in a sticky mucus layer that the cilia find difficult to move. This sticky mucus in the lungs has two major effects on health. It increases the chances of lung infection and it makes gas exchange less efficient, particularly in the later stages of the disease.

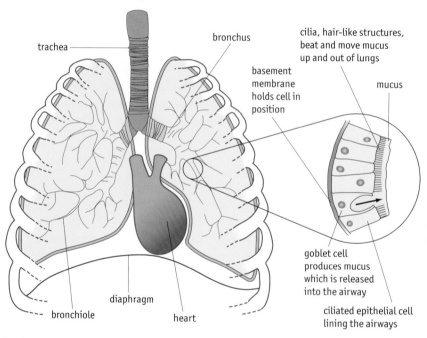

Figure 2.2 The fine structure of the lungs. Cilia, hair-like extensions from the surface of cells lining the trachea, bronchi and bronchioles, have an important role in keeping the lungs clean. By a repeated beating motion they move mucus and particles up and out of the lungs.

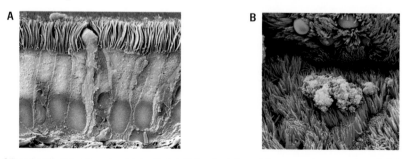

Figure 2.3 A Ciliated epithelial cells and goblet cells (pink) line the tubes of the gas exchange system. **B** False-colour scanning electron micrograph of a clump of mucus (green/orange) surrounded by cilia (purple) in the lungs of someone with CF. Rounded goblet cells (light blue) secrete the mucus.

What are epithelial cells?

Epithelial cells form the outer surface of many animals including mammals. They also line the cavities and tubes within the animal and cover the surfaces of internal organs. The cells work together as a **tissue** known as an **epithelium** (plural epithelia). In Topic 3 you will study how the cells of multicellular organisms can be organised into tissues.

The epithelium consists of one or more layers of cells sitting on a **basement membrane**. This is made of protein fibres in a jelly-like protein-carbohydrate matrix. This basement membrane layer anchors the epithelium to the connective tissue below. The membrane surface of the epithelial cell that faces the basement membrane is known as its basal membrane and the surface that faces away from it is the apical membrane. There are several different types of epithelia.

The epithelium in the walls of the alveoli and capillaries is squamous (or pavement) epithelium. The very thin flattened cells fit together like crazy paving. The cells can be less than 0.2 μm thick. The apical membrane faces the lumen.

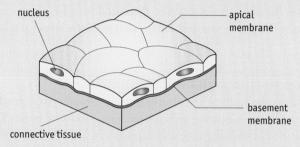

Squamous or pavement epithelium.

In the small intestine the epithelial cells extend out from the basement membrane. These column-shaped cells make up columnar epithelium. The free surface facing the intestine lumen is normally covered in microvilli, which greatly increase the surface area for absorption.

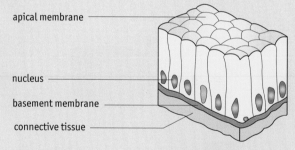

Columnar epithelium.

In the trachea, bronchi and bronchioles there are ciliated epithelial cells with cilia (hair-like structures) on the apical surface. These cilia beat in a coordinated way like a Mexican wave and move substances along the tube they line. The ciliated columnar epithelium of the airways in the lungs appears to be stratified (composed of several layers), but in fact each cell is in contact with the basement membrane. It appears to have several layers because some cells have their nucleus at the base of the cell while in others it is in the centre, giving the impression of different layers. This epithelium is therefore known as pseudostratified.

Ciliated epithelium.

CF problems

How sticky mucus increases the chances of lung infections

Microorganisms become trapped in the mucus in the lungs and some of these microorganisms can cause illness – they are **pathogens**. The mucus is normally moved by cilia into the back of the mouth cavity where it is either coughed out or swallowed, reducing the risk of infection. Acid in the stomach kills most microorganisms that are swallowed.

In people with CF the mucus layer is so sticky that the cilia cannot move it. Mucus production still continues, as it would in a normal lung, and the layers of thickened mucus build up in the airways. There are low levels of oxygen in the mucus, partly because oxygen diffuses slowly through it, and partly because the epithelial cells use up more oxygen in CF patients. Harmful bacteria can thrive in these anaerobic conditions.

Q 2.1 Why does ineffective removal of mucus create a problem for people with CF?

Q 2.2 Why does swallowing mucus reduce the risk of infection?

White blood cells fight the infections within the mucus but as they die they break down and release DNA that makes the mucus even stickier. Repeated lung infections can eventually weaken the body's ability to fight the pathogens and cause damage to the structures of the gas exchange system.

How sticky mucus reduces gas exchange

Gases such as oxygen cross the walls of the alveoli into the blood system by **diffusion**. To supply enough oxygen to all the body's respiring cells gas exchange must be rapid. To understand the impact of sticky mucus on gas exchange we must appreciate how the fine structure of the lungs helps to maximise this process.

KEY BIOLOGICAL PRINCIPLE: THE EFFECT OF INCREASE IN SIZE ON SURFACE AREA

Living organisms have to exchange substances with their surroundings. For example, they take in oxygen and nutrients and get rid of waste materials such as carbon dioxide. In unicellular organisms the whole cell surface membrane is the exchange surface. Substances that diffuse into or out of a cell move down a concentration gradient (from a high to a low concentration). The gradients are maintained by the cell continuously using the substances absorbed and producing waste. For example, oxygen diffusing into a cell is used for respiration, which produces carbon dioxide.

The larger an organism, the more exchange has to take place to meet the organism's metabolic needs. Larger multicellular organisms have more problems absorbing substances because of the size of the organism's surface area compared with its volume. This is known as the surface area-to-volume ratio, calculated by dividing an organism's total surface area by its volume.

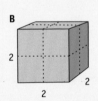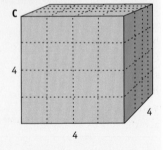

Q 2.3 For each of the 'organisms' in the next column, work out its surface area, volume and then surface area-to-volume ratio. The units are arbitrary.

Q 2.4 As the organism doubles in size what happens, quantitatively, to **a** its surface area, **b** its volume, **c** its surface area-to-volume ratio?

Q 2.5 Assuming that this organism relies on diffusion across its outer surface for exchange, why would it have problems if it grew any larger?

Q 2.6 If you compared a tiger, a horse and a hippopotamus, which would have the smallest surface area-to-volume ratio?

It is clear that as organisms get larger, the surface area per unit of volume gets less. If larger organisms relied only on their general body surface for exchange of substances they could not survive because the distance to the innermost tissues is too far for diffusion to supply oxygen quickly enough; exchange would be too slow.

> **⚙ ACTIVITY**
> Complete **Student Activity 2.3** to investigate the effect of surface area-to-volume ratio on uptake by diffusion.

How can an organism increase in volume while still managing to exchange enough nutrients by diffusion?

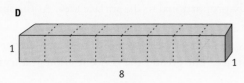

Q 2.7 Work out the surface area, volume and surface area-to-volume ratio of organisms D and E.

Q 2.8 Look at the values you have calculated for B, D and E. What do you notice?

Q 2.9 If a slug, an earthworm and a tapeworm all had the same volume, which would have the largest surface area-to-volume ratio?

All three blocks, B, D and E, have the same volume but they have very different surface areas. The most elongated block, D, has the largest surface area-to-volume ratio.

Relying on the outer body surface for gas exchange is only possible in organisms with a very small volume, or in larger organisms that have a high enough surface area-to-volume ratio such as worms with a tubular or flattened shape.

Q 2.10 If a land-living organism were to use its entire external surface covering for gas exchange, what problems might it encounter in addition to slow diffusion?

In larger organisms a variety of special organs have evolved that increase the surface area for exchange, therefore increasing the surface area-to-volume ratio. For example, a mammal's lungs provide a large surface for gas exchange while minimising heat and water loss from the moist surface. Digestive systems provide a large surface area for food absorption.

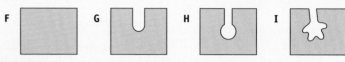

Q 2.11 Which of the four organisms above, all with the same volume, has the largest surface area for exchange?

Q 2.12 Name two organs that enable rapid exchange.

Q 2.13 In humans, how do the substances that are absorbed get to all the distant parts of the body?

Gas exchange surfaces

Within the lungs, alveoli provide a large surface area for exchange of gases between the air and the blood. Figure 2.4 is a cross-section of the alveoli. Imagine filling a box with balloons and then slicing through it with a cheese slicer. One slice would look like this stained section. Figure 2.5 shows the alveoli in three dimensions. Look at these two figures and identify features of the **gas exchange surface** that you think would ensure rapid exchange between air in the alveoli and the blood.

You should have noticed some of these features of the gas exchange surface:

● large surface area of the alveoli

● numerous capillaries around the alveoli

● thin walls of the alveoli and capillaries meaning a short distance between the alveolar air and blood in the capillaries.

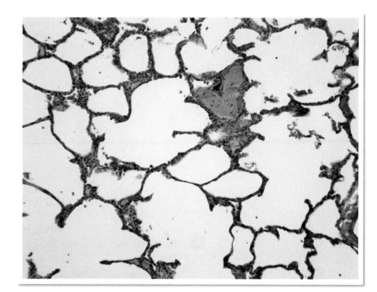

Figure 2.4 Ventilation of the lungs ensures that the air in the alveoli is frequently refreshed. This helps maintain a steep concentration gradient and maximise gas exchange across the walls of the alveoli.

ACTIVITY

In **Student Activity 2.4** you will examine slides of the alveoli to observe and measure the features that aid diffusion into and out of the bloodstream.

In the web tutorial, **Student Activity 2.5**, you can investigate the surface area of the lungs.

The body's demand for oxygen is enormous so diffusion across the alveolar wall needs to be rapid. The rate of diffusion is dependent on three properties of gas exchange surfaces.

● **Surface area** – rate of diffusion is directly proportional to the surface area. As the surface area increases the rate of diffusion increases.

● **Concentration gradient** – rate of diffusion is directly proportional to the difference in concentration across the gas exchange surface. The greater the concentration gradient the faster the diffusion.

● **Thickness of the gas exchange surface** – rate of diffusion is inversely proportional to the thickness of the gas exchange surface. The thicker the surface the slower the diffusion.

Combining all these factors, we can state that

$$\text{rate of diffusion} \propto \frac{\text{surface area} \times \text{difference in concentration}}{\text{thickness of the gas exchange surface}}$$

This relationship is known as **Fick's law**.

The large surface area of the alveoli, the steep concentration gradient between the alveolar air and the blood (maintained by ventilation of the alveoli and the continuous flow of blood through the lungs), and the thin walls of the alveoli and capillaries, combine to ensure rapid diffusion across the gas exchange surface (Figure 2.5).

CHECKPOINT

2.1 Describe the properties of gas exchange surfaces.

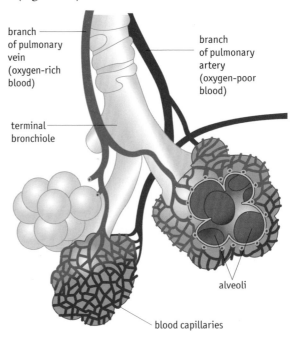

Figure 2.5 The structure of the alveoli and the surrounding capillaries ensures that there is rapid diffusion across the gas exchange surface.

How sticky mucus affects gas exchange

The sticky mucus layer in the bronchioles of a person with cystic fibrosis tends to block these narrow airways, preventing movement of air into the alveoli beyond the blockage. This reduces the number of alveoli providing surface area for gas exchange. Blockages are more likely at the narrow ends of the airways. These blockages will often allow air to pass when the person breathes in but not when they breathe out, resulting in over-inflation of the lung tissue beyond the blockage. This can damage the elasticity of the lungs.

People with CF sometimes find it difficult to take part in physical exercise because their gas exchange system cannot deliver enough oxygen to their muscle cells. The oxygen is needed for the chemical processes of aerobic respiration that release the energy used to drive the contraction of the muscles during exercise. People with CF become short of breath when taking exercise although exercise is very beneficial to them.

2.2 Why is CF mucus so sticky?

In people with CF, the mucus layer on the surface of the epithelial cells is sticky because it contains less water than normal. The reduced water level is due to abnormal salt (sodium chloride) and water transport across the cell surface membranes caused by a faulty transport protein channel in the membrane. To understand what is going on you need to be clear about the structure and functions of proteins, the structure of cell surface membranes and how substances are transported in and out of cells. You also need to be aware of how proteins are made so you can understand what has gone wrong with the protein channel in someone with CF.

KEY BIOLOGICAL PRINCIPLE: PROTEIN STRUCTURE IS THE KEY TO PROTEIN FUNCTION

Proteins have a wide range of functions in living things. Antibodies, enzymes and many hormones are all protein molecules. Various proteins make up muscles, ligaments, tendons and hair. Proteins are also components of cell membranes and have important functions within the membrane. All proteins are composed of the same basic units: amino acids. Proteins may typically contain between 50 and 2000 amino acids. There are 20 different amino acids that occur commonly in proteins. Plants can make all these amino acids but animals can only make some, obtaining the others through their diet. The amino acids that animals have to obtain in their diet are known as essential amino acids.

Figure 2.6A shows the general structure of an amino acid. All amino acids have a common structure. In every amino acid a central carbon atom is bonded to an amine group ($-NH_2$), a carboxylic acid group ($-COOH$), a hydrogen ($-H$) and a residual group ($-R$). Each different amino acid has a different R group. Glycine is the simplest amino acid with a hydrogen ($-H$) forming its R group (Figure 2.6B). Alanine's R group is a methyl ($-CH_3$) group. Some amino acids have R groups with more complex carbon ring structures. R groups may also contain nitrogen atoms or a sulphur atom, for example, $- CH_2 CH_2CONH_2$ in glutamine and $-CH_2SH$ in cysteine.

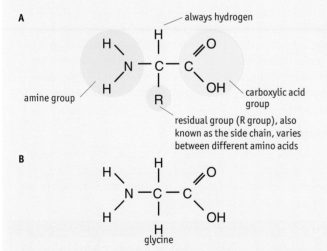

Figure 2.6 A The general structure of amino acids. **B** The structure of glycine.

Primary structure

Two amino acids join in a condensation reaction to form a dipeptide, with a **peptide bond** forming between the two subunits, as shown in Figure 2.7. This process can be repeated to form polypeptide chains which may contain thousands of amino acids. A protein is made up of one or more of these polypeptide chains. The sequence of amino acids in the polypeptide chains is known as the **primary structure** of a protein.

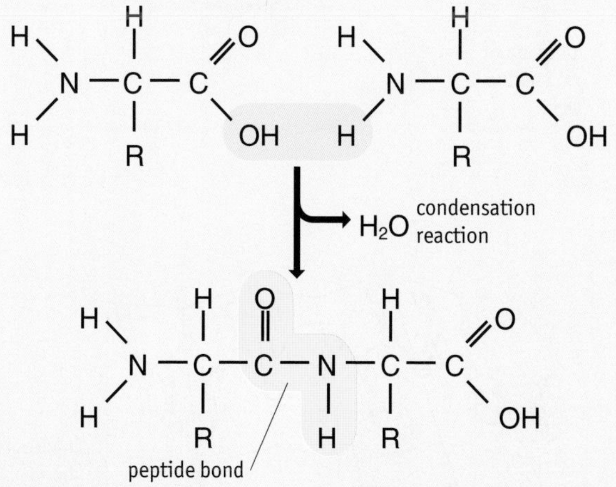

Figure 2.7 Two amino acids link in a condensation reaction to form a dipeptide. Numerous amino acids link together in this way to form a polypeptide.

Further levels of protein structure

Interactions between the amino acids in the polypeptide chain cause the chain to twist and fold into a three-dimensional shape. Lengths of the chain may first coil into α-helices or come together in β-pleated sheets (Figure 2.8). These features are known as the **secondary structure**. The chain then folds into its final three-dimensional shape, the **tertiary structure** (Figure 2.9).

Secondary structure

The chain of amino acids may twist to form an α-helix, a shape like an extended spring (Figure 2.8A). Within the helix, hydrogen bonds form between the slightly negative C=O of the carboxylic acid and the slightly positive $-NH$ of the amine group of different amino acids that lie above and below each other, stabilising the shape. Sections of α-helix can be up to about 35 amino acids long.

Amino acid chains may fold back on themselves, or several lengths of the chain, each up to about 15 amino acids in length, may link together with hydrogen bonds holding the parallel chains in an arrangement known as a β-pleated sheet (Figure 2.8B). Each hydrogen bond is weak but the cumulative effect of many H bonds makes the secondary structure quite stable. Within one protein molecule there may be sections with α-helices and other sections that contain β-pleated sheets, and some sections may be folded or twisted in a less ordered manner.

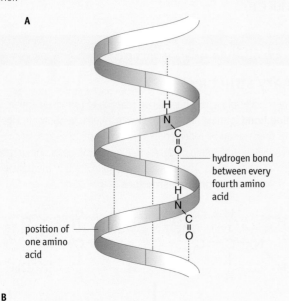

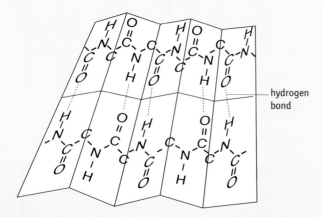

Figure 2.8 The secondary structure of proteins. **A** An α-helix. **B** A β-pleated sheet (R groups are not shown).

> ### SUPPORT
> To find out more about hydrogen bonding, have a look at the Biochemistry support on the website.

Tertiary and quaternary structure

A polypeptide chain often bends and folds further to produce a precise three-dimensional shape. Chemical bonds and hydrophobic interactions between R groups maintain this final tertiary structure of the protein (Figure 2.9).

An R group is polar when the sharing of the electrons within it is not quite even. Polar R groups attract other polar molecules, like water, and are therefore hydrophilic (water-attracting). The non-polar groups are hydrophobic (water-repelling). Non-polar hydrophobic R groups are arranged so they face the inside of the protein, excluding water from the centre of the molecule. Chemical bonds may form between R groups that are close to each other in the folded structure. The amino acid cysteine has an R group that contains an –SH group. If two cysteine R groups are close to each other a covalent disulphide bond may form. Some amino acids contain R groups that are ionised (charged) and therefore ionic bonds can form between positively and negatively charged R groups. Disulphide and ionic bonds are much stronger than the hydrogen bonds but are very sensitive to changes in pH.

If the three dimensional structure is functional, that is, the molecule is able to perform its specific function, the molecule is now described as a protein.

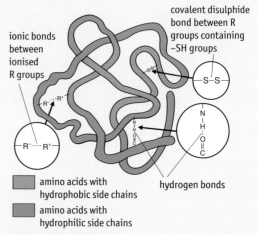

Figure 2.9 The three-dimensional structure of a protein is held in place by chemical bonds between individual amino acids and by hydrophobic interactions between R groups.

Some proteins may only be functional if they are made up of several polypeptide chains held together. For example, haemoglobin, the protein found in red blood cells that carries oxygen, is made up of four polypeptide chains held tightly together in a structure known as the quaternary structure. Only proteins with more than one polypeptide chain have a quaternary structure; single chain proteins stop at the tertiary level.

Conjugated proteins

Some proteins are known as conjugated proteins. They have another chemical group associated with their polypeptide chain(s). For example, the polypeptide chains that make up myoglobin and haemoglobin are associated with an iron-containing group.

> ### ACTIVITY
> The interactive tutorial in **Student Activity 2.6** lets you review how amino acids join to form a polypeptide and then fold to achieve their three-dimensional structure.

Globular and fibrous proteins

Proteins can be divided into two distinct groups according to their overall shapes:

● globular proteins
● fibrous proteins.

In **globular proteins** the polypeptide chain is folded into a compact spherical shape. These proteins are soluble due to the hydrophilic side chains that project from the outside of the molecules and are therefore important in metabolic reactions. Enzymes are globular proteins. Their three-dimensional shape is crucial to their ability to form enzyme-substrate complexes and catalyse reactions within cells.

The three-dimensional shapes of globular proteins are critical to their roles in binding to other substances. Examples include transport proteins within membranes and the oxygen-transport pigments haemoglobin (in red blood cells) and myoglobin (in muscle cells), seen in Figure 2.10. Antibodies are also globular and rely on their precise shapes to bind to the microorganisms that enter our bodies.

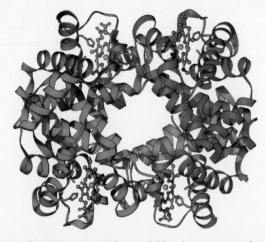

Figure 2.10 The globular protein haemoglobin acts as an oxygen transport molecule. It is made up of four polypeptide chains, each is associated with an iron-containing haem group (grey). Oxygen attaches to the iron with the haem group. Because the protein is associated with another chemical group it is called a conjugated protein.

Fibrous proteins do not fold up into a ball shape but remain as long chains. Several polypeptide chains can be cross-linked for additional strength. These insoluble proteins are important structural molecules. Keratin in hair and skin, and collagen (Figure 2.11) in the skin, tendons, bones, cartilage and blood vessel walls, are examples of fibrous proteins.

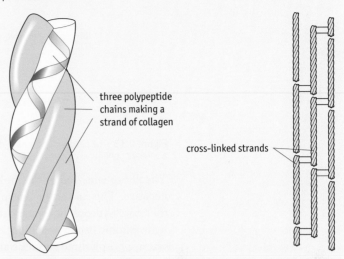

three polypeptide chains making a strand of collagen

cross-linked strands

Figure 2.11 Collagen is a fibrous protein. Three polypeptide chains wind around each other to form a rope-like strand held together by hydrogen bonds between the chains. Each strand cross-links to other strands to produce a molecule with tremendous strength. Notice that the strands are staggered, avoiding the creation of any weak points along the length of the molecule.

> **✓ CHECKPOINT**
>
> **2.2** Compare and contrast the structures and functions of globular and fibrous proteins.

Cell membrane structure

A phospholipid bilayer

CF is caused by a faulty transport protein in the surface membranes of epithelial cells. But how do we know what the structure of the cell surface membrane is like? It is so thin that under a light microscope a cross-section of the membrane looks like a single line. However, closer examination using an electron microscope reveals that it is in fact a **bilayer**, about 7 nm wide, appearing as two distinct lines (Figure 2.12). The basic structure is two layers of **phospholipids**.

Look back at Figure 1.39 (page 34) and remind yourself of the structure of a triglyceride lipid molecule – three fatty acids and a glycerol. Compare this with Figure 2.13 and notice the difference. In a phospholipid there are only two fatty acids; a negatively charged phosphate group replaces the third fatty acid.

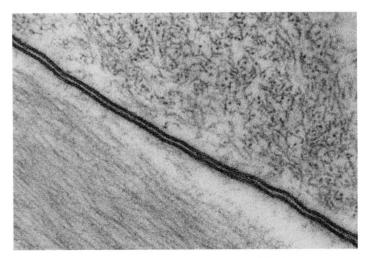

Figure 2.12 The cell surface membrane appears as two distinct layers when viewed with an electron microscope. The same type of membrane surrounds many cell organelles, which you will study in Topic 3.

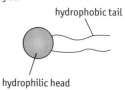

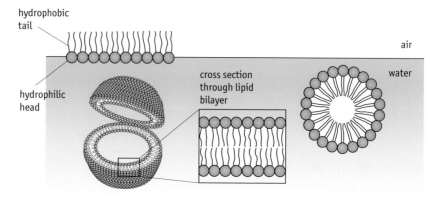

Figure 2.13 The phospholipid molecule has two distinct sections: the hydrophilic head and the hydrophobic tails. X can be a variety of chemical groups..

The phosphate head of the molecule is polar; one end is slightly positive and the rest is slightly negative. This makes the phosphate head attract other polar molecules, like water, and it is therefore hydrophilic (water-attracting). The fatty acid tails are non-polar and therefore hydrophobic (water-repelling). We all know that fats and water do not mix. When added to water, phospholipids become arranged with no contact between the hydrophobic tails and the water. They may form a layer on the surface with their hydrophobic tails directed out of the water. Alternatively they may become arranged into spherical clusters called micelles or form a bilayer (Figure 2.14). The bilayer is favoured by phospholipids because the two fatty acids are too bulky to fit into the interior of a micelle. The formation of bilayers by phospholipids is of critical biological importance. A lipid bilayer will tend to close on itself so that there are no ends with exposed hydrocarbon chains, thus forming compartments as happens around and within cells.

SUPPORT

To find out more about polar molecules and polar bonds, have a look at the Biochemistry support on the website.

Figure 2.14 Phospholipids in water may form a monolayer on the surface, spherical micelles, or bilayers.

Cells are filled with an aqueous (watery) cytoplasm and are surrounded by aqueous tissue fluid. The cell surface-membrane phospholipids tend to adopt the most stable arrangement, a bilayer (Figure 2.15). In this arrangement the hydrophobic fatty acid tails have no contact with the water on either side of the membrane and the hydrophilic phosphate heads remain in contact with the aqueous environment.

The fluid mosaic model

The cell surface membrane is not simply a phospholipid bilayer. It also contains proteins, cholesterol, **glycoproteins** (protein molecules with polysaccharides attached) and **glycolipids** (lipid molecules with polysaccharides attached). Some of the proteins span the membrane. Other proteins are found only within the inner layer or only within the outer layer (Figure 2.15). Membrane proteins have hydrophobic areas and these are positioned within the membrane bilayer.

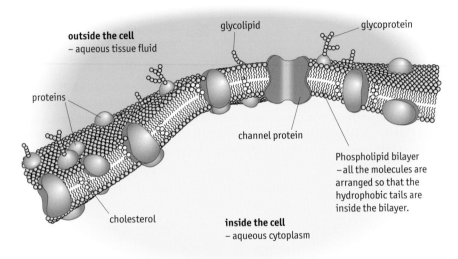

Figure 2.15　Diagram of the fluid mosaic model of the cell surface membrane.

It is thought that some of the proteins are fixed within the membrane, but others are not and can move around in the fluid phospholipid bilayer. This arrangement is known as the **fluid mosaic model** of membrane structure.

Two Californian biochemists, S. Jonathan Singer and Garth Nicholson, first proposed the fluid mosaic model in 1972. They did not carry out all the experiments that supported the model themselves, but were able to bring together evidence obtained by a number of different research groups and incorporate it into their proposal.

Evidence for the fluid mosaic model

The most widely accepted membrane model up until the early 1970s was a three-layer protein-lipid sandwich, based on the evidence from electron micrographs in which the dark outer layers were thought to be proteins and the lighter region within was thought to be the lipid (Figure 2.16A). However, the protein-lipid sandwich model does not allow the hydrophilic phosphate heads to be in contact with water, nor does it allow any non-polar hydrophobic amino acids on the outside of the membrane proteins to be kept away from water. Consideration of how lipids behaved in water, forming a bilayer because it is the most stable arrangement, was used to refine the model. Interpretation of the electron micrograph evidence changed to support the new model of the membrane structure. The phosphate heads are more electron dense and show up as the darker edges to the membrane with the lipid tails being the lighter inner part.

Experiments showed that there were two types of proteins – those that could be dissociated (separated) from the membrane quite easily by increasing the ionic strength of the solution, and those (the majority) that could only be removed from the membrane by more drastic action such as adding detergents. This evidence supported the fluid mosaic model where some **peripheral proteins** are loosely attached on the outside surface of the membrane whilst **integral proteins** are fully embedded within the phospholipids, some even spanning both layers. Several integral proteins were investigated further and shown to have regions at their ends that had polar hydrophilic amino acids, with the middle portion being mainly composed of non-polar hydrophobic amino acids (Figure 2.16B).

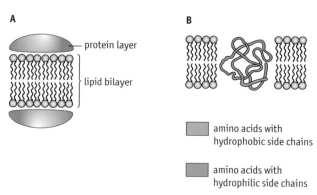

Figure 2.16 **A** The protein-lipid sandwich model. **B** In the fluid mosaic model integral membrane proteins have polar and non-polar regions.

Additional evidence for integral membrane proteins came from freeze-fracture electron microscopy studies. Frozen membrane sections were fractured along the weak point between the lipid layers (Figure 2.17) and the inner fractured surface coated in a heavy metal. Scanning electron microscopy, which gives three-dimensional images, revealed a smooth mosaic-like surface (the lipid tails) interspersed by much larger particles (the integral proteins) (Figure 2.18).

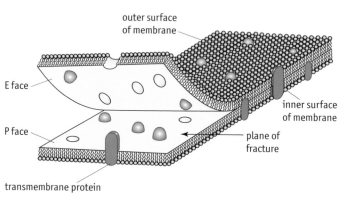

Figure 2.17   Freeze-fracture of membranes reveals the integral transmembrane proteins.

Figure 2.18 Scanning electron micrograph showing a freeze fracture preparation of epithelial cell membrane. The lower half of the micrograph shows the P face (see Figure 2.17), the upper half is the E face. The bumps are integral proteins that extend across the membrane. The rougher surface on the face closer to the cytoplasm side of the membrane shows that more proteins are anchored on this side.

Finally, several experiments were carried out using labelled molecules that only attach to other specific molecules. In one experiment, plant proteins called lectins that bind to polysaccharides were labelled with ferritin (a protein with a core of ferric oxide) which is visible under the electron microscope. When mixed with membrane samples the lectins only bound to the outer surface of the membrane, and never to the inner. This showed that membranes are asymmetric, that is, the outside surface of the membrane is different to the inside, and once again did not support the protein sandwich model, which would have had lectin binding on both sides.

Another neat experiment involved fusing mouse cells with human cells. Before the cells were fused, a specific membrane protein was labelled in each cell type. The mouse membrane protein was given a green fluorescent label, and the human membrane protein was given a red fluorescent label. A light microscope was used to follow where the green and red proteins moved. Immediately after fusing the cells, the coloured labels remained in their respective halves, but after 40 minutes at 37 °C there was complete intermixing of the proteins (Figure 2.19). The only way the proteins could have intermixed was to have diffused through the membranes, showing the components were indeed fluid.

Q 2.14   When the mouse and human membrane protein experiment was carried out at 15 °C, mixing was much slower. How does this evidence further support the fluid mosaic model of membrane structure?

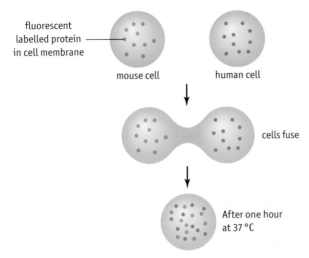

fluorescent labelled protein in cell membrane

mouse cell   human cell

cells fuse

After one hour at 37 °C

Figure 2.19 The movement of the fluorescent labelled proteins within the cell surface membranes supported the fluid mosaic model of membrane structure.

More unsaturated phospholipids – more fluid

The greater the ratio of phospholipids that contain unsaturated fatty acids to those containing saturated fatty acids, the more fluid the membrane will be. The 'kinks' in the hydrocarbon tails of the unsaturated phospholipids prevent them from packing closely together, so more movement is possible (Figure 1.45 and 2.20). Cholesterol sits between the phospholipids and maintains the fluidity of the membrane by affecting movement of the phospholipids (Figure 2.15).

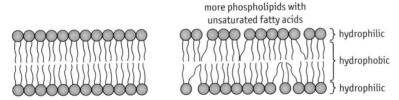

more phospholipids with unsaturated fatty acids

hydrophilic

hydrophobic

hydrophilic

Figure 2.20 More phospholipids with unsaturated fatty acids will make the membrane more fluid.

Q 2.15   Can you suggest why the membrane is more fluid with unsaturated rather than saturated phospholipids making up the bilayer?

Many different types of protein are found within membranes, each type having a specific function. Some proteins function as enzymes, others function as carrier and channel proteins involved in the transport of substances in and out of cells. Glycoproteins and glycolipids have important roles in cell-to-cell recognition and as receptors.

How do substances pass through cell membranes?

For a cell to function correctly it needs to be able to control transport across its surface membrane. Molecules and ions move across membranes by:

- diffusion
- osmosis
- active transport
- exocytosis
- endocytosis.

> **ACTIVITY**
>
> In **Student Activity 2.7** evaluate the evidence for different models that explain the structure and function of cell membranes. You can investigate the cell surface membrane practically in **Student Activity 2.8**.

Diffusion

Diffusion is the net (overall) movement of molecules or ions from a region where they are at a higher concentration to a region of their lower concentration. Particles are continually moving randomly in all directions but where there is a high concentration of particles there is increased probability that the particles move away towards the area of lower concentration, resulting in the overall *net* movement. Diffusion will always occur where there is a difference in concentration between two areas, known as a concentration gradient. Diffusion will continue until equilibrium when the particles of the substance are evenly spread throughout the whole volume. Small uncharged particles can diffuse directly across the cell membrane, passing between the lipid molecules as they move down a concentration gradient. Small molecules, including oxygen and carbon dioxide, diffuse in this way, moving rapidly across the cell membrane. Carbon dioxide is polar but its small size still allows rapid diffusion.

Facilitated diffusion

Hydrophilic (polar) molecules and ions that are larger than carbon dioxide cannot simply diffuse through the bilayer. They are insoluble in lipids – the hydrophobic tails of the phospholipids provide an impenetrable barrier. Instead they cross the membrane with the aid of proteins in a process called **facilitated diffusion**. The polar molecules and ions may diffuse through water-filled pores within **channel proteins** that span the membrane (Figure 2.15). There are different channel proteins for transporting different molecules. Each type of channel protein has a specific shape that permits the passage of only one particular type of ion or molecule. Some channels can be opened or closed depending on the presence or absence of a signal, which could be a specific molecule, like a hormone, or a change in **potential difference** (voltage) across the membrane. These channels are called **gated channels**.

Some proteins that play a role in facilitated diffusion are not just simple channels but are **carrier proteins**. The ion or molecule binds onto a specific site on the protein. The protein changes shape (Figure 2.21) and as a result the ion or molecule crosses the membrane. The movement can occur in either direction, with the net movement being dependent on the concentration difference across the membrane. Molecules move from high to low concentration due to more frequent binding to carrier proteins on the side of the membrane where the concentration is higher.

Diffusion, whether facilitated or not, is sometimes called **passive transport**. 'Passive' here refers to the fact that no metabolic energy is needed for the transport – the process is driven by the concentration gradient itself.

Osmosis

Osmosis is the net movement of water molecules from a solution with a lower concentration of solute to a solution with a higher concentration of solute through a partially permeable membrane. Osmosis is summarised in Figure 2.22.

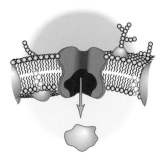
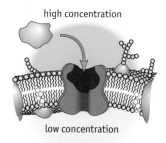

high concentration

low concentration

Figure 2.21 The carrier protein changes shape, facilitating diffusion.

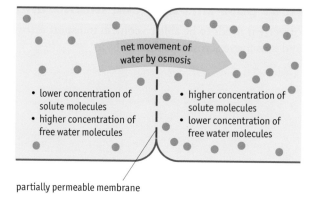

net movement of water by osmosis

• lower concentration of solute molecules
• higher concentration of free water molecules

• higher concentration of solute molecules
• lower concentration of free water molecules

partially permeable membrane

Figure 2.22 Osmosis summary.

> **ACTIVITY**
>
> **Student Activity 2.9** lets you investigate different methods of transport.

Look at the left and right parts of Figure 2.23 below. In each case decide the direction in which the solvent (i.e. water) molecules will have a net tendency to move. Only the right half of Figure 2.23 has a partially permeable membrane so osmosis can only take place there. Osmosis is due to the random movement of water molecules across the membrane and is a particular type of diffusion. If solute molecules are present, water molecules form hydrogen bonds with them, and this reduces the movement of these water molecules. If more solute is present there are fewer free water molecules able to collide with and move across the membrane. Osmosis will continue until the solutions on either side are equally concentrated or **isotonic**.

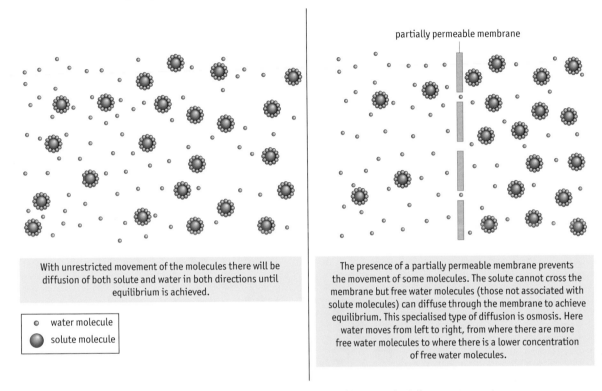

partially permeable membrane

With unrestricted movement of the molecules there will be diffusion of both solute and water in both directions until equilibrium is achieved.

The presence of a partially permeable membrane prevents the movement of some molecules. The solute cannot cross the membrane but free water molecules (those not associated with solute molecules) can diffuse through the membrane to achieve equilibrium. This specialised type of diffusion is osmosis. Here water moves from left to right, from where there are more free water molecules to where there is a lower concentration of free water molecules.

- ● water molecule
- ● solute molecule

Figure 2.23 Decide which direction the water molecules will move in, then read the shaded boxes to check if you were correct.

Active transport

If substances need to be moved across a membrane against a concentration gradient (from low concentration to high concentration) then energy is required. As with facilitated diffusion, specific carrier proteins are also needed. The energy is supplied by the energy transfer molecule **ATP** (adenosine triphosphate). ATP provides the immediate source of energy for all biological processes. It is formed during respiration, the breakdown of energy storage molecules, principally carbohydrates and fats. The substance to be transported across the membrane binds to the carrier protein. One phosphate group is removed from ATP by hydrolysis and ADP (adenosine diphosphate) forms. A small amount of energy is required to break the bond holding the end phosphate in the ATP. Once removed, the phosphate group becomes hydrated. A lot of energy is released as bonds form between water and phosphate. This energy from ATP changes the shape of the carrier protein, causing the substance to be released on the other side of the membrane and moving it against the concentration gradient (Figure 2.24).

Active transport proteins are sometimes referred to as pumps, and the pumping of substances across membranes occurs in every cell. Examples appear throughout the course, including transport of ions across epithelial cells (later in Topic 2), plant cell roots (Topic 4), muscle cells (Topic 6), and nerve cells (Topic 8). Active transport also occurs between compartments within a cell, for example between the mitochondria and cytoplasm (Topics 3 and 7).

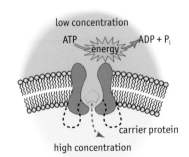

low concentration

ATP   energy   ADP + P_i

carrier protein

high concentration

Figure 2.24 In active transport energy is required to change the shape of the carrier protein (dotted line) and move the substance across the membrane against a concentration gradient.

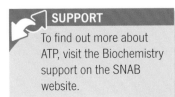

SUPPORT

To find out more about ATP, visit the Biochemistry support on the SNAB website.

Exocytosis and endocytosis

Sometimes very large molecules or particles, or very large quantities of a particular molecule need to be transported across cell surface membranes. This **bulk transport** is achieved by **exocytosis** and **endocytosis**, which rely on the fluid nature of the membrane (Figure 2.25). Exocytosis is the release of substances, usually proteins or polysaccharides, from the cell. Vesicles (small membrane-bound sacs containing the substance) fuse with the cell membrane and the contents are released. For example, insulin (the hormone produced by certain cells in the pancreas) is released into the blood by exocytosis. Neurotransmitter substances are also released in this way from nerve endings (Topic 8). Endocytosis is the reverse process. Substances are taken into a cell by the creation of a vesicle from the cell surface membrane. Part of the cell membrane engulfs the solid or liquid material to be transported. In some cases the substance to be absorbed attaches to a receptor in the membrane and is then absorbed by endocytosis. This is how cholesterol is taken up into cells. White blood cells ingest bacteria and other foreign particles by endocytosis (Topic 6).

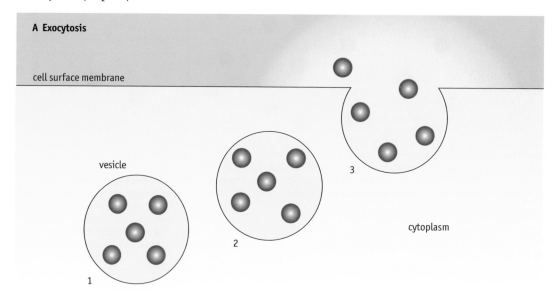

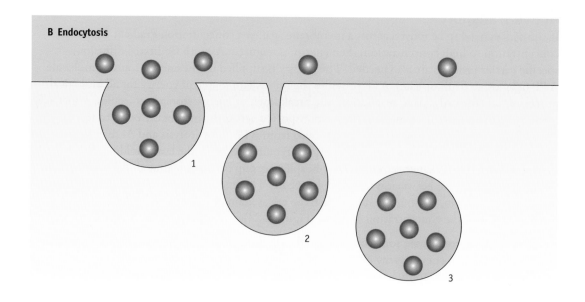

Figure 2.25 In exocytosis, vesicles fuse with the cell surface membrane, releasing their contents from the cell. In endocytosis, substances are brought inside a cell within vesicles formed from the cell surface membrane.

Table 2.1 provides a summary of the different methods of transport across cell membranes.

Diffusion	down a concentration gradient, from high to low concentration of the substance until equilibrium reached
	hydrophobic (lipid-soluble) or small uncharged molecules
	through phospholipid bilayer
	passive, no energy required
Facilitated diffusion	down a concentration gradient, from high to low concentration of the substance until equilibrium reached
	hydrophilic molecules or ions
	through channel proteins or via carrier proteins that change shape
	passive, no energy required
Osmosis	a type of diffusion involving movement of free water molecules
	from high to low concentration of free water molecules until equilibrium reached (free water molecules move from region of low solute concentration to one of high concentration of solute)
	through phospholipid bilayer
	passive, no energy required
Active transport	against a concentration gradient, low to high concentration
	through carrier proteins that change shape
	requires energy, supplied by ATP
Exocytosis	used for bulk transport of substances out of the cell
	vesicles fuse with the cell surface membrane, releasing their contents
Endocytosis	used for bulk transport of substances into the cell
	vesicles are created from the cell surface membrane, bringing their contents into the cell

Table 2.1 Summary of methods of transport across cell membranes.

Q 2.16   For each example below, suggest the type of transport most likely to be involved.

(a) Movement of oxygen across the wall of an alveolus.

(b) Absorption of phosphate ions into root hair cells.

(c) Pumping of calcium ions into storage vesicles inside muscle cells.

(d) Release of glucose from liver cells into the bloodstream.

(e) Removal of the sodium ions that diffuse into a nerve cell, to maintain a low concentration within the nerve axon.

(f) Reabsorption of water molecules from the kidney tubule.

What happens in the membranes of the cells lining the airways?

The cells that line the airways produce mucus. In people who do not have cystic fibrosis, the water content of the mucus is continuously regulated to maintain a constant **viscosity** ('stickiness') of the mucus. It must be runny enough to be moved by the beating cilia but not so runny that the fluid floods the airway. This regulation of the water content of the mucus is achieved by the transport of sodium ions and chloride ions across the epithelial cells. Water then follows the ions by osmosis.

Comparative studies of the ionic composition of mucus in CF and normal mice, and in CF and healthy humans have provided evidence supporting the following theory that explains what is happening in the lungs of people with CF.

Regulating mucus water content in unaffected lungs

Excess water in the mucus

If the mucus layer is too runny, the presence of excess water is detected by the epithelial cell membranes that line the airways. Carrier proteins in the basal membranes of the epithelial cells actively pump sodium ions out of the cells (Figure 2.26). The concentration of sodium ions (Na^+) in the cell falls, setting up a concentration gradient across the apical membrane. Sodium ions diffuse from the mucus down this concentration gradient into the epithelial cell by facilitated diffusion through epithelial sodium ion channels (ENaCs) in the apical membrane.

The raised concentration of Na^+ in the tissue fluid on the basal membrane side of the epithelial cells creates a potential difference between this tissue fluid and the mucus on the apical membrane side. The tissue fluid now contains more positively charged ions than does the mucus. This creates an electrical gradient between the tissue fluid and the mucus. This electrical gradient causes negatively charged chloride ions (Cl^-) to diffuse out of the mucus into the tissue fluid via the gaps between neighbouring epithelial cells.

The overall effect of these processes is to increase the Na^+ and Cl^- concentrations in the tissue fluid. Water is then drawn out of the epithelial cells by osmosis across the basal membrane into the tissue fluid. This water loss increases the overall solute concentration within the cell. Since the solute concentration is now higher within the cell than in the mucus, water is drawn out of the mucus by osmosis across the apical membrane and into the epithelial cell.

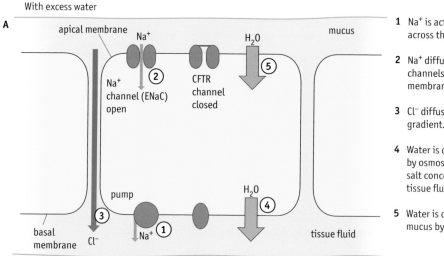

Figure 2.26 The role of Na^+ and Cl^- movement and osmosis in removing excess water from the mucus to prevent it from becoming too runny.

Having too much water in the mucus is a normal state of affairs. This is because the cilia are continuously moving mucus along the airways. Movement of mucus from numerous smaller bronchiole branches into fewer larger bronchioles means that water must be removed to reduce the volume of mucus so as to avoid the larger airways flooding with fluid.

Too little water in the mucus

When there is too little water in the mucus, for example after a period of rapid breathing during exercise, chloride ions are transported across the basal membrane into the epithelial cell (Figure 2.27). This creates a concentration gradient across the apical membrane, with the concentration of chloride ions being higher inside the cell than out. In a person who does not have CF, this chloride ion imbalance causes the cystic fibrosis transmembrane conductance regulator **(CFTR)** protein channels to open. The CFTR protein is a type of gated channel protein. Chloride ions now diffuse out of the cell through the CFTR channels down this concentration gradient into the mucus. When open, the CFTR channels block (close) the epithelial sodium ion channels (ENaC) in the apical membrane. The exact mechanism for this is currently unknown but may include other proteins. The build up of negatively charged chloride ions in the mucus creates an electrical gradient between the mucus and the tissue fluid. Sodium ions diffuse out of the tissue fluid and move down this electrical gradient, passing between the cells into the mucus. The movement of the sodium and chloride ions into the mucus draws water out of the epithelial cells by osmosis until the solutions are isotonic. This movement of water prevents the mucus that lines the airways from becoming too viscous (sticky).

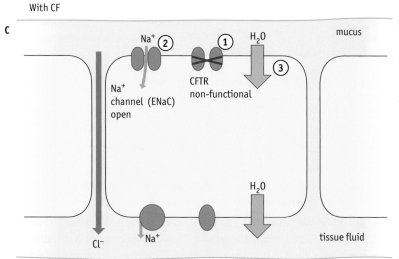

With too little water

1 Cl⁻ is pumped into the cell across the basal membrane.

2 Cl⁻ diffuses through the open CFTR channels.

3 Na⁺ diffuses down the electrical gradient into the mucus.

4 Elevated salt concentration in the mucus draws water out of the cell by osmosis.

5 Water is drawn into the cell by osmosis.

Figure 2.27 The role of the CFTR channel in preventing mucus becoming too viscous.

Why CF lungs cannot regulate the water in mucus

In a person with CF, the CFTR protein may be missing, or if it is present it does not function correctly (Figure 2.28). When there is too little water in the mucus and it is sticky, chloride ions cannot be secreted across the apical membrane. The epithelial sodium ion channels (ENaCs) are not blocked and actually seem to allow even more sodium ions than normal into the epithelial cells. Since the ENaCs are always open, there is continual sodium absorption from the mucus by the epithelial cells. The raised levels of sodium ions in the cells then draw chloride ions and water out of the mucus into the cells by osmosis. This makes the mucus even more viscous which makes it harder for the beating cilia to move the mucus. The mucus is not effectively cleared up and out of the lungs and this build-up reduces the effective ventilation of the alveoli. The mucus frequently becomes infected with bacteria, and phagocytic cells that clear pathogens are over-produced in response. When the phagocytes break down, their DNA makes the mucus stickier still, causing a downward spiral of airway inflammation and lung damage.

With CF

1 CFTR channel is absent or not functional.

2 Na⁺ channel is permanently open.

3 Water is continually removed from mucus by osmosis.

Figure 2.28 The effect of non-functioning CFTR channel on mucus viscosity.

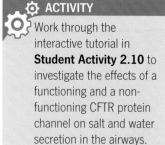

ACTIVITY
Work through the interactive tutorial in **Student Activity 2.10** to investigate the effects of a functioning and a non-functioning CFTR protein channel on salt and water secretion in the airways.

2.3 How does cystic fibrosis affect other body systems?

The effect of CF on the digestive system

Cystic fibrosis sufferers have difficulty maintaining body mass because of problems with the digestion and absorption of nutrients. They also have high basal metabolic rates. They generally have poor appetites but still have to eat more than most people, including high-energy food to make sure they obtain sufficient nutrients and energy. They require 120–140% of the recommended daily energy intake. People with CF may also take food supplements that contain digestive enzymes. The aim of these supplements is to help break down large food molecules.

Most of the chemical breakdown of food molecules and the subsequent absorption of the soluble products into the bloodstream occurs in the small intestine. Glands secrete digestive enzymes into the lumen of the gut, where they act as catalysts to speed up the extracellular breakdown of food molecules. Exocrine glands outside the gut, for example, in the salivary glands, liver and pancreas, produce a wide range of enzymes (Figure 2.29). Enzymes are also built into the membranes of the gut wall.

Groups of pancreatic cells produce enzymes that help in the breakdown of proteins, carbohydrates and lipids. These digestive enzymes are delivered to the gut in pancreatic juice released through the pancreatic duct (Figure 2.30). The pancreas is also an endocrine gland. Different groups of pancreatic cells release hormones, including insulin which is involved in regulating blood sugar levels.

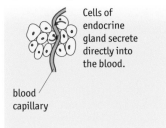

A Exocrine gland

Cells of exocrine gland secrete into duct.

duct carries secretions to where they are required

surface epithelium

B Endocrine gland

Cells of endocrine gland secrete directly into the blood.

blood capillary

Figure 2.29 There are two types of gland. These are: **A** exocrine glands with ducts and **B** endocrine glands which are ductless. Endocrine glands release hormones.

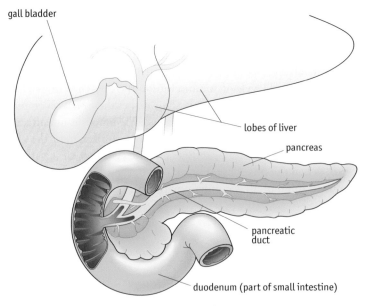

gall bladder

lobes of liver

pancreas

pancreatic duct

duodenum (part of small intestine)

Figure 2.30 Thick mucus can easily block the pancreatic duct in a CF patient.

In a person with CF, the pancreatic duct becomes blocked by sticky mucus, impairing the release of digestive enzymes. The lower concentration of enzymes within the small intestine reduces the rate of digestion. Food is not fully digested, so not all the nutrients can be absorbed. As a consequence the faeces contain a higher proportion of partially digested and undigested food, so energy is lost. This is called malabsorption syndrome.

An additional complication occurs when the pancreatic enzymes become trapped behind the mucus blocking the pancreatic duct. These enzymes damage the pancreas itself. The cysts of hard, damaged or fibrosed tissue within the pancreas give the CF condition its name. Another complication occurs if damage occurs to cells within the pancreas that produce the hormone insulin, which is involved in the control of blood sugar levels. A form of diabetes can be the result.

CHECKPOINT

2.5 For each of **a** to **c**, write five bullet point statements to describe the effect CF has on: **a** the gas exchange system **b** the digestive system **c** the reproductive system.

KEY BIOLOGICAL PRINCIPLE: ENZYME FUNCTION DEPENDS ON PROTEIN THREE-DIMENSIONAL STRUCTURE

Enzymes are globular proteins that act as biological **catalysts**. They speed up chemical reactions that would otherwise occur very slowly at the temperature within cells. Usually, every reaction within the cell is catalysed by a specific enzyme. The precise three-dimensional shape adopted by an enzyme includes a depression on the surface of the molecule called the **active site**. The active site is the part of the enzyme molecule with the catalytic function. The site may be a relatively small part of the large protein molecule, as you can see in Figure 2.31. Only a few amino acids may be directly involved in the active site, with the remainder maintaining the three-dimensional shape of the protein molecule. The active site of the enzyme will have a particular shape. There are several theories that explain how the active site has a catalytic effect.

Lock-and-key theory

Either a single molecule with a complementary shape, or more than one molecule that together have a complementary shape, can fit into the active site (Figure 2.32). These **substrate** molecule(s) form temporary bonds with the amino acids of the active site to produce an **enzyme-substrate complex**. The enzyme holds the substrate molecule(s) in such a way that they react more easily. When the reaction has taken place the products are released, leaving the enzyme unchanged. The substrate is often likened to a 'key' which fits into the enzyme's 'lock', so this is known as the **lock-and-key theory** of enzyme action. Each enzyme will only catalyse one specific reaction because only one shape of substrate will fit into its precisely-shaped active site.

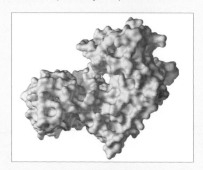

Figure 2.31 The active site is only a small part of the enzyme molecule. Here, the active site of the enzyme hexokinase changes shape as the glucose (yellow) enters.

Induced fit theory

It has been found that the active site is often flexible. When the substrate (or substrates) enters the active site, the enzyme molecule changes shape slightly, fitting more closely around the substrate. It is like a person putting on a wetsuit; the wetsuit shape changes to fit the body but returns to its original shape when taken off. This is known as the **induced fit theory** of enzyme action. Only a specifically shaped substrate will induce the correct change in shape of an enzyme's active site. The slight shape change of the active site enables the substrates to react.

Activation energy

To convert substrate(s) into product(s), bonds must change both within and between molecules. Breaking chemical bonds requires energy, while energy is released when bonds form. The energy needed to break bonds and start the reaction is known as the **activation energy**. Without an enzyme, heating a substrate would provide this energy. Think about starting a bonfire – a reaction between the molecules in wood and oxygen. You must first provide some energy to start the fire. The heat energy agitates atoms within the molecules, the molecules become unstable and the reaction can then proceed. In cells, enzymes reduce the amount of energy needed to bring about a reaction; this allows reactions to occur without raising the temperature of the cell.

How do enzymes reduce the activation energy?

The specific shape of the enzyme's active site and of its complementary substrate(s) is such that electrically charged groups on their surfaces interact (Figure 2.32). The attraction of oppositely charged groups may distort the shape of the substrate(s) and assist in the breaking of bonds or in the formation of new bonds. In some cases, the active site may contain amino acids with acidic side chains and the acidic environment created within the active site may provide conditions favourable for the reaction.

Q 2.17 Figure 2.32 shows the splitting of a substrate molecule into two products. Describe how this diagram would need to be changed to show the formation of a single product molecule from two substrate molecules.

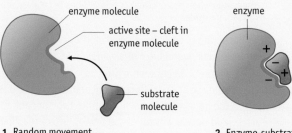

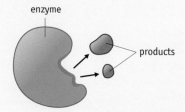

1 Random movement causes the enzyme and substrate to collide, and the substrate enters the active site.

2 Enzyme-substrate complex forms. Charged groups attract, distorting the substrate and aiding bond breakage or cristation.

3 Products are released from the active site leaving the enzyme unchanged and ready to accept another substrate molecule.

Figure 2.32 Charged side groups interact and assist in breaking or forming bonds.

Enzymes are present in all organisms and catalyse a huge range of reactions. They speed up reactions by at least a million times and most biological reactions would not happen at all without enzymes. The metabolism of an organism is the sum of all the enzyme-catalysed reactions occurring within it. Some metabolic reactions are extremely simple, like the hydration of carbon dioxide, which is catalysed by carbonic anhydrase:

$$\text{Carbonic anhydrase}$$
$$CO_2 + H_2O \rightleftharpoons H_2CO_3$$

This reaction allows carbon dioxide to be transported in the blood from respiring tissues to the alveoli.

Enzymes catalyse many different types of reactions, some occurring inside cells (**intracellular reactions**) and others in tissue fluid, blood or other aqueous solutions in the organism (**extracellular reactions**). Inside cells, enzymes can be free floating or attached to membranes. Sometimes large substrate molecules will be broken down into smaller units in **catabolic** (breaking down) reactions. The hydrolysis of starch to maltose by salivary amylase is an extracellular catabolic reaction. Other enzymes catalyse **anabolic** (building up) reactions, like the fatty acid synthetase enzyme complex that, unsurprisingly, synthesises fatty acids within cells. Sometimes substrate molecules are simply altered, not added to or broken down – isomerase enzymes do this.

Q 2.18 For each of the enzymes below decide if they are extracellular or intracellular, and if they are catalysing catabolic or anabolic reactions.

Enzyme	Reaction catalysed
decarboxylase	removal of carboxyl group (–COOH) in respiration with the formation of carbon dioxide
maltase	breakdown of maltose to glucose in digestion
DNA polymerase	joins nucleotides together in DNA replication
catalase	breakdown of hydrogen peroxide (H_2O_2), a toxic by-product of metabolism
pancreatic lipase	breakdown of triglycerides into glycerol and fatty acids

✓ CHECKPOINT

2.3 Write a definition for each of the following key enzyme terms:

- biological catalyst
- activation energy
- active site
- enzyme-substrate complex.
- product
- lock-and-key theory
- induced fit theory

2.4 Explain how the three-dimensional structures of proteins enable enzymes to perform their functions as biological catalysts.

As we have seen, enzymes:

- are globular proteins
- have an active site that allows binding with a specific substrate
- catalyse (speed up) reactions (intracellular and extracellular)
- reduce the activation energy required for a chemical reaction to take place
- do not alter the end-product or nature of a reaction
- do not get used up and remain unchanged at the end of a reaction and able to bind with another substrate molecule.

Finding rates of enzyme-controlled reactions

The rate of reaction is measured by determining the quantity of substrate used or the quantity of product formed in a given time. For example, when the enzyme catalase is used to break down hydrogen peroxide (H_2O_2) to water and oxygen, the rate of reaction can be found by measuring the volume of oxygen given off in a known time.

If we mix a fixed quantity of enzyme and substrate, at first the reaction will proceed quickly, as shown in Figure 2.33. However, as the substrate is used up, there are fewer substrate molecules to collide with and bind to the enzyme so the reaction slows down and eventually stops and no further increase in the product occurs, as seen in Figure 2.33. The slope of the rapid phase of the reaction is known as the initial rate of reaction and is frequently used when comparing rates of enzyme-controlled reactions. A steeper slope means a higher initial rate of reaction.

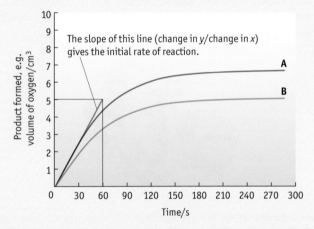

Figure 2.33 The quantity of product is measured over time to determine the progress of an enzyme-catalysed reaction. Here the volume of oxygen produced is measured when two different concentrations of catalase are added to hydrogen peroxide.

The initial rate of reaction is calculated by dividing the change in the y axis by the change in the x axis values.

Q 2.19 Figure 2.33 shows the progress of two reactions catalysed by catalase at two different concentrations, A and B. Compare the initial rate of reaction for the two concentrations of enzyme and decide which has the slower rate of reaction. Support your answer with rates calculated from the graph.

↪ SUPPORT

To find out more about calculating rate of change see the Maths support sheet 6 – rates of change on the website.

How do enzyme and substrate concentrations affect the rate of reaction?

Figure 2.34A shows the effect of enzyme concentration on the initial rate of reaction. The initial rate of reaction is directly proportional to the enzyme concentration because the more enzyme that is present, the greater the number of active sites that are available to form enzyme-substrate complexes. The increase in rate will continue in this linear fashion assuming that there is an excess of substrate. In Figure 2.34B

you can see that at high substrate concentrations it is the enzyme concentration that limits the rate of reaction. Every active site is occupied and substrate molecules cannot enter an active site until one becomes free again.

> ⚙ **ACTIVITY**
>
> In **Student Activity 2.11** you can investigate the effect of enzyme concentration on enzyme activity.

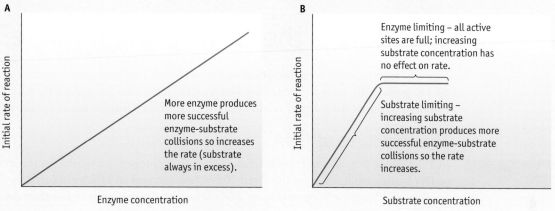

Figure 2.34 The effect of **A** enzyme concentration and **B** substrate concentration on the rate of an enzyme-controlled reaction.

The effect of CF on the reproductive system

We have seen how sticky mucus produced by a defective CFTR protein can lead to major complications in the lungs and pancreas. In the reproductive system it can also cause severe problems. Females have a reduced chance of becoming pregnant because a mucus plug develops in the cervix (Figure 2.35A). This stops sperm from reaching the egg. Males with cystic fibrosis commonly lack the vas deferens (sperm duct) on both sides, which means that sperm cannot leave the testes (Figure 2.35B). Where the vas deferens is present it can become partially blocked by a thick sticky mucus layer. This means fewer sperm are present in each ejaculate.

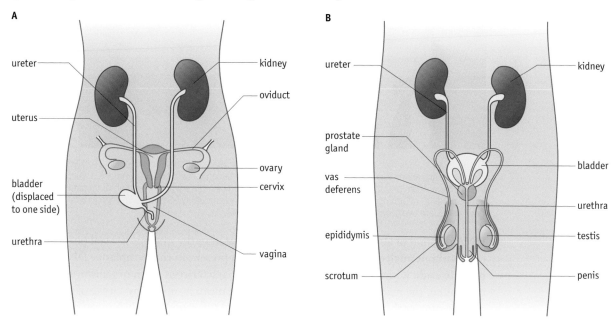

Figure 2.35 A The female reproductive system. In CF a mucus plug develops in the cervix. **B** The male reproductive system. In CF the vas deferens may be absent or plugged with mucus.

The effect of CF on sweat

Sweat glands are exocrine glands that initially secrete into their lumen a solution of salt and water that is isotonic to the blood. In an individual without cystic fibrosis, CFTR and ENaC proteins allow reabsorption of sodium chloride from the sweat as it moves up the duct towards the skin surface. The sweat that is released on to the skin and evaporates is therefore **hypotonic**.

Q 2.20 Unusually salty sweat is often one of the first signs that a baby may have CF. Why might the sweat of a person with CF be more salty than normal?

2.4 How is the CFTR protein made?

Cystic fibrosis is caused by a **mutation** in the DNA that carries the instructions for making the CFTR protein. In order to know how DNA works (and what has gone wrong in cystic fibrosis), you need to know what a gene is and how the codes it contains are used to make proteins.

The structure of DNA

In 1953 James Watson and Francis Crick (Figure 2.36) proposed a model for the structure of DNA using evidence from the X-ray diffraction patterns of DNA produced by Rosalind Franklin and Maurice Wilkins (Figure 2.37). Their model was correct and their discovery has revolutionised biology.

WEBLINK

You can find out how the X-ray diffraction pattern was created and how it was interpreted by visiting the DNA interactive website section called Code: Finding the Structure.

EXTENSION

You can read about the controversy surrounding the work of Rosalind Franklin and DNA in **Student Extension 2.1**.

Figure 2.36 James Watson (left) and Francis Crick (right) worked out the structure of DNA in 1953. 'The seed of life itself. Peel the chains apart, each chain reproduces the other, one becomes two, two become one. Generation on generation, all the way from Adam and Eve to you and me. It never dies. One simple shape. The womb of humanity. Endlessly, effortlessly fertile, dividing, reforming… It's the closest we'll ever get to immortality.' *Source:* Tim Piggott-Smith, who played Francis Crick in the film *Life Story*, talking about the structure of DNA.

DNA is found in almost every cell. DNA is located in the nucleus if a cell possesses one. In cells without nuclei (e.g. bacterial cells) DNA is present in the cytoplasm. DNA contains the **genetic code** which dictates all the inherited characteristics of an organism. DNA does this by controlling the manufacture of proteins. An organism is unique by virtue of the DNA it contains and the sum total of all the proteins it produces. Differences in DNA mean different proteins are produced which means organisms differ from each other. Organisms of different species show bigger differences between their DNA and proteins than organisms of the same species. DNA carries the genetic information from one generation to the next.

A   **B**

 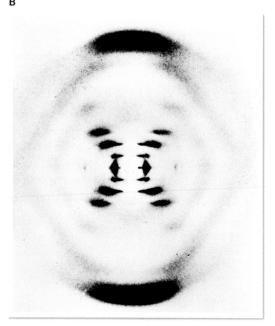

Figure 2.37 **A** Rosalind Franklin. **B** The X-ray diffraction photos of crystallised DNA obtained by Rosalind Franklin and Raymond Gosling in 1952. The 'X' pattern is produced by the helical structure of DNA; using measurements from the X-ray, Franklin and Wilkins worked out the dimensions of the DNA structure.

Gene and genome

A **gene** is a sequence of bases on a DNA molecule that codes for a sequence of amino acids in a polypeptide chain. Each chromosome found in the cell contains a large amount of DNA and carries numerous genes. Genes make up only a fraction of the total length of DNA in the chromosomes. Some of the rest of the DNA is involved in regulating or controlling the production of proteins but the job of the remainder of the DNA is not fully known. Together, all the genes in an individual (or species) are known as the **genome**.

DNA is a chain of nucleotides

DNA is one type of nucleic acid, called **deoxyribonucleic acid**. It is a long chain polymer made of many units called **nucleotides** or **mononucleotides**.

Q 2.21   DNA is called a polynucleotide. Explain why.

A mononucleotide contains three molecules linked together by condensation reactions. They are **deoxyribose** (a 5-carbon sugar), a **phosphate group** and an **organic base** containing nitrogen. Look at Figure 2.38 to see how these three are arranged in a mononucleotide.

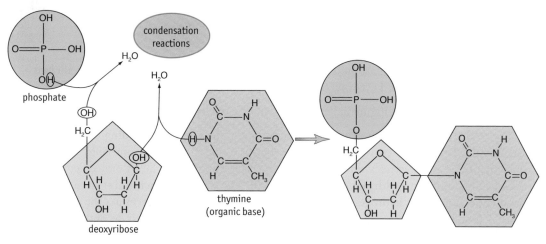

Figure 2.38 A deoxyribose sugar, a phosphate and a base join to form a mononucleotide (often known as a nucleotide).

The nitrogen-containing base is the only part of the nucleotide that is variable. There are four possible bases: **adenine**, **cytosine**, **guanine** and **thymine**. These bases are often represented by their initial letters, **A**, **C**, **G** and **T**, respectively.

Mononucleotides link together by condensation reactions between the sugar of one nucleotide and the phosphate of the next one, producing a polynucleotide. The bond that forms between the two nucleotides is known as a **phosphodiester bond**.

In a DNA molecule there are two polynucleotide strands twisted around each other to form a double helix (Figure 2.39), rather like a spiral staircase or a twisted ladder. The sugars and phosphates form the two sugar-phosphate 'backbones' of the molecule and are on the outside. The bases point inwards horizontally and are held together in pairs by hydrogen bonds. The DNA in each human cell contains some 3000 million of these base pairs. The two nucleotide strands are described as being **antiparallel** because they run in opposite directions.

SUPPORT

To remind yourself about hydrogen bonds, visit the Biochemistry support on the website.

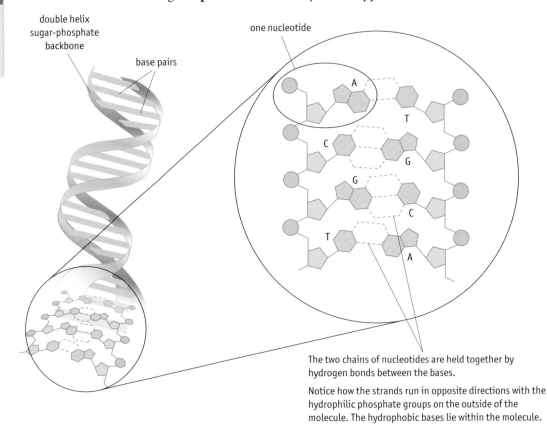

The two chains of nucleotides are held together by hydrogen bonds between the bases.

Notice how the strands run in opposite directions with the hydrophilic phosphate groups on the outside of the molecule. The hydrophobic bases lie within the molecule.

Figure 2.39 The structure of DNA. Remember that any picture you see of DNA is a simplification. It is a model based on evidence from techniques such as X-ray diffraction. A whole DNA molecule is much longer than this and usually contains millions of base pairs.

Why do the bases pair up?

If you look at Figure 2.39 you should notice that the bases only pair in a certain way: adenine only pairs with thymine, and cytosine only pairs with guanine.

Q 2.22 Look very closely at Figure 2.39 and suggest why the bases might only form these pairs.

The key to this pairing is in the structure of the bases and the hydrogen bonding between them. Bases A and G both have a two-ring structure, whereas C and T only have one ring. The bases pair so that there are effectively three rings forming each 'rung' of the DNA molecule, making the molecule a uniform width along its whole length. The shape and chemical structure of the bases dictates how many hydrogen bonds each one can form and this determines the pairing of A with T (two hydrogen bonds) and C with G (three hydrogen bonds). This deceptively simple fact is the clue to how DNA works. The bases A and T and the bases C and G are referred to as **complementary base pairs**.

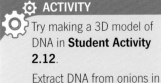**ACTIVITY**

Try making a 3D model of DNA in **Student Activity 2.12**.

Extract DNA from onions in **Student Activity 2.13**.

Q 2.23 **(a)** The sequence of bases on part of one strand of a DNA molecule is:

A T C C C T G A G G T C A G T

What would be the sequence of bases on the corresponding part of the other strand?

(b) Analysis of a quantity of DNA revealed 54% to be composed of A-T base pairs. What proportion of the DNA is cytosine?

How does DNA code for proteins?

The CF gene is on chromosome 7. It is a long gene, made up of about 230 kbp (230 000 base pairs). It instructs the cell to make the CFTR protein that forms the transmembrane chloride channel. The sequence of bases in the DNA is the genetic code that tells the cell which amino acids to link together to make the CFTR protein. Every gene is a sequence of bases on a DNA molecule coding for a sequence of amino acids in a polypeptide chain. The polypeptide chain then twists and folds into a functional protein.

From DNA to proteins

DNA, and the genetic code it contains, is in the nucleus, but proteins are made in the cytoplasm. DNA cannot pass through the membranes surrounding the nucleus into the cytoplasm. How does the genetic code get from the nucleus to the cytoplasm, and how are the proteins assembled in the cytoplasm? How does the sequence of bases get converted into a sequence of amino acids? In other words, how does **protein synthesis** occur?

Protein synthesis

The first stage of protein synthesis is called **transcription**. Transcription takes place in the nucleus. A molecule that is a 'copy' of the gene coding for the required protein is made. This 'copy' is not made from DNA but from another type of nucleic acid called **ribonucleic acid (RNA)**. This RNA molecule can leave the nucleus, carrying the information to the cytoplasm where it is used in the manufacture of proteins. The original gene on the DNA stays in the nucleus. This is like making a photocopy of some precious original plans so that everyone on the factory floor can work from them, leaving the originals safe at head office.

What is the difference between DNA and RNA?

An RNA molecule is a *single* stranded polynucleotide made of **ribonucleic acid (RNA) nucleotides**. RNA nucleotides are very similar in structure to DNA nucleotides, containing a phosphate, sugar and base, except that they contain **ribose** sugar and not deoxyribose (Figure 2.40). Another difference is that in RNA nucleotides, the base **uracil (U)** replaces thymine, so RNA *never* contains thymine. Sometimes a section of an RNA molecule folds back on itself and complementary bases pair with each other, so it can appear double stranded, although it is not.

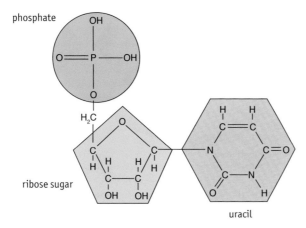

Figure 2.40 Compare this RNA nucleotide with the DNA nucleotide in Figure 2.38 and notice the similarities and differences.

The RNA made in the nucleus during transcription is known is known as **messenger RNA (mRNA)** because it carries the code from the DNA to the cytoplasm where it is used in the manufacture of proteins. The correct sequence of amino acids join together in the second stage of protein synthesis called **translation** (Figure 2.41). Translation involves two other types of RNA: **transfer RNA (tRNA)** and **ribosomal RNA (rRNA)**. The two stages of protein synthesis are described in detail below.

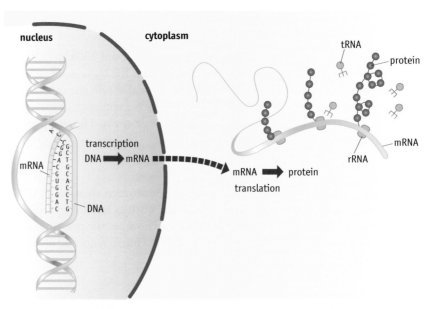

Figure 2.41 In transcription, the DNA code is copied onto mRNA. In translation, it is used to join the correct sequence of amino acids and build a new protein.

Protein synthesis – transcription

At the start of transcription, an enzyme called **RNA polymerase** attaches to the DNA (Figure 2.42). The hydrogen bonds between paired bases break, and the DNA molecule unwinds. The sequence on one of the strands, **the template strand**, is transcribed to makes an mRNA molecule with the same base sequence as the DNA coding strand. The complementary RNA nucleotides align themselves into position and then phosphodiester bonds form to produce an mRNA molecule. Because of complementary base pairing, the order of bases on the DNA exactly determines the order of the bases on the mRNA. Only the section of DNA that codes for the

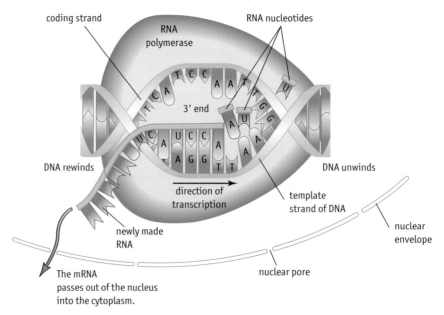

Figure 2.42 Protein synthesis – transcription.

protein being made is transcribed. When transcription is complete, the mRNA molecule leaves the nucleus through a pore in the **nuclear envelope** and the DNA molecule 'zips up'.

The DNA coding strand is also known as the sense strand, while the template strand is also known as the antisense strand, because once transcribed it makes an mRNA molecule with the same base sequence as the DNA coding (sense) strand. Be aware that many biology text books are incorrect in their use of the terms sense strand and antisense strand.

Protein synthesis – translation
In normal English, translation is the process of turning words from one language into another. The biological process of translation means turning the sequence of bases in the genetic code (the language of nucleotides) into a sequence of amino acids (the language of proteins). In a cryptic code, a letter, word or symbol represents or stands for something else – the genetic code works in much the same way.

The nature of the genetic code
In the genetic code, one base does not simply code for one amino acid. There are only four bases, so if this was the case proteins could contain only four different amino acids, instead of the 20 amino acids commonly found. The code carried by the DNA is a three-base or **triplet code** (Figure 2.43). Each adjacent group of three bases codes for an amino acid. The code is **non–overlapping**, each triplet code is adjacent. There are 64 possible three-letter combinations if the four DNA bases are grouped in triplets. One triplet sequence is a start code and three are stop codes (called chain terminators). This leaves on average three different triplet codes for each of the 20 naturally occurring amino acids in protein molecules. Because several triplets can code for the same amino acid the code is described as **degenerate**. This amazingly simple but fundamental coding system is universal, it is found in all organisms.

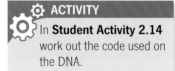
ACTIVITY
In **Student Activity 2.14** work out the code used on the DNA.

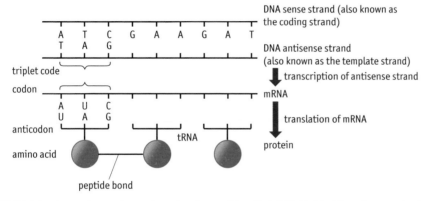

Figure 2.43 Triplet codes, codons and complementary base pairing. Each DNA triplet of three bases codes for one amino acid. Several triplets code for the same amino acids.

During transcription every triplet code on DNA gives rise to a complementary **codon** on mRNA (Figure 2.43). Table 2.2 overleaf shows all the mRNA codons and their corresponding amino acids. Notice that in some cases, all the codes with the same first two letters code for the same amino acid. For instance, all triplet codes starting GU code for valine. On the other hand, a couple of amino acids (tryptophan and methionine) are coded for by a single triplet code. You do not need to learn the triplet codes or codons of the genetic code.

Q 2.24 Look at Figure 2.43. State which bases are missing from the unlabelled triplet codes on the antisense strand and the unlabelled codons on the mRNA.

A change or **mutation** in the sequence of bases in DNA could change a triplet that makes up a gene. This could then change the amino acid sequence of the primary structure which may substantially alter the protein's three-dimensional structure and its properties. Mutations in DNA include substitutions, insertions, deletions and inversions of base sequences. Insertions and deletions of a number of bases where the number cannot be divided by three causes a 'frame shift'. All the subsequent triplets from that point onwards are affected.

mRNA codons		Second base							
		U		C		A		G	
First base	U	UUU	Phe	UCU	Ser	UAU	Tyr	UGU	Cys	U
		UUC	Phe	UCC	Ser	UAC	Tyr	UGC	Cys	C
		UUA	Leu	UCA	Ser	UAA	Stop	UGA	Stop	A
		UUG	Leu	UCG	Ser	UAG	Stop	UGG	Trp	G
	C	CUU	Leu	CCU	Pro	CAU	His	CGU	Arg	U
		CUC	Leu	CCC	Pro	CAC	His	CGC	Arg	C
		CUA	Leu	CCA	Pro	CAA	Gln	CGA	Arg	A
		CUG	Leu	CCG	Pro	CAG	Gln	CGG	Arg	G
	A	AUU	Ile	ACU	Thr	AAU	Asn	AGU	Ser	U
		AUC	Ile	ACC	Thr	AAC	Asn	AGC	Ser	C
		AUA	Ile	ACA	Thr	AAA	Lys	AGA	Arg	A
		AUG	Met	ACG	Thr	AAG	Lys	AGG	Arg	G
	G	GUU	Val	GCU	Ala	GAU	Asp	GGU	Gly	U
		GUC	Val	GCC	Ala	GAC	Asp	GGC	Gly	C
		GUA	Val	GCA	Ala	GAA	Glu	GGA	Gly	A
		GUG	Val	GCG	Ala	GAG	Glu	GGG	Gly	G

Key:

Ala = alanine	Gly = glycine	Pro = proline
Arg = arginine	His = histidine	Ser = serine
Asn = asparagine	Ile = isoleucine	Thr = threonine
Asp = aspartic acid	Leu = leucine	Trp = tryptophan
Cys = cysteine	Lys = lysine	Tyr = tyrosine
Gln = glutamine	Met = methionine	Val = valine
Glu = glutamic acid	Phe = phyenylalanine	

Table 2.2 A look-up table showing which mRNA codons code for each amino acid. The amino acids' names are always shortened to three letters, as in this table.

Translation of the code

Once in the cytoplasm, mRNA attaches to a ribosome where it causes translation to take place and amino acids are linked together in the correct order according to the sequence of codons on the mRNA. Ribosomes are small organelles (structures within the cell) made of ribosomal RNA and protein. Ribosomes are found free in the cytoplasm or attached to endoplasmic reticulum, a system of flattened, membrane-bound sacs. Ribosomes are composed of two subunits, a smaller and a larger subunit. The larger subunit contains two tRNA binding sites. The mRNA attaches to the smaller subunit, so that two mRNA codons face the two binding sites of the larger subunit (Figure 2.44).

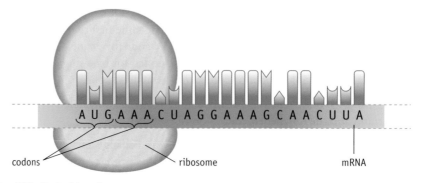

Figure 2.44 mRNA attached to a ribosome.

Transfer RNA (tRNA) molecules are the key to translation, they are 'bilingual'. Each amino acid has its own tRNA molecule that transfers amino acid present in the cytoplasm to the ribosome.

Although made of single stranded RNA, tRNA molecules fold back on themselves and complementary base pairing makes them a clover leaf shape. In diagrams tRNA molecules can be represented in several ways (Figure 2.45). At one end of the tRNA is a three base sequence called an **anticodon**. At the opposite end of the tRNA is a binding site for an amino acid.

The anticodon sequence is complementary to the mRNA codon for a particular amino acid (Figure 2.46). For example, the mRNA codons for the amino acid lysine are AAA and AAG. The complementary anticodons on tRNAs are UUU and UUC. Within the cytoplasm, free amino acids become attached to the correct tRNA molecules. Each amino acid has its own specific tRNA which carries it to the ribosome. See Table 2.2 for the full genetic code.

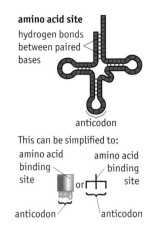

Figure 2.45 The tRNA molecule carries an amino acid. Its anticodon base pairs with the codon on the mRNA molecule.

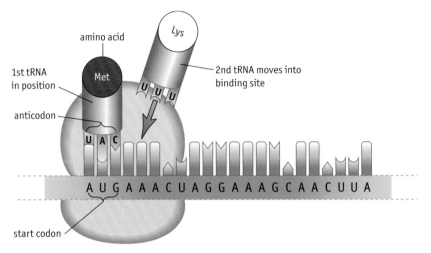

Figure 2.46 tRNA molecules complementary base pair with mRNA.

The first codon exposed on the ribosome is always the start codon AUG. This is the code for the amino acid methionine. The tRNA molecule carrying methionine has the complementary anticodon, UAC, and it now hydrogen bonds to the start codon AUG. The next codon is facing the next binding site. This codon attracts the tRNA-amino acid complex with the complementary anticodon, and it binds to it.

The ribosome holds the mRNA, tRNAs, amino acids and associated enzyme in place while a **peptide bond** forms between the two amino acids (Figure 2.47). The peptide bond forms in a condensation reaction between the amine group of one amino acid and the carboxylic acid group of the next, forming a dipeptide.

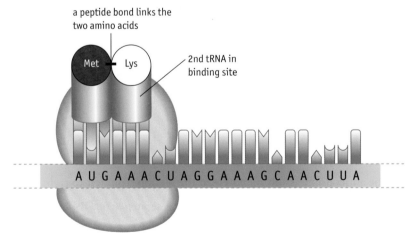

Figure 2.47 A peptide bond forms between the two amino acid molecules.

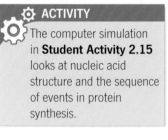
Once the peptide bond has formed, the ribosome moves along the mRNA to reveal a new codon at the binding site. The first tRNA returns to the cytoplasm where it can collect another methionine (Figure 2.48). tRNA molecules do not get used up but are continuously ferrying amino acids from the cytoplasm to the ribosome to be incorporated into polypeptides. Translation continues with different tRNAs transferring the correct amino acids to the growing polypeptide on the ribosome, and condensation reactions forming peptide bonds, until a stop codon (chain terminator) is reached. The stop codons are UAA, UAG or UGA. There are no tRNAs with complementary anticodons to these sequences so no amino acid can be transferred, the polypeptide chain stops growing and detaches from the ribosome. In Topic 3 you will discover how and where the protein completes its folding and becomes fully functional.

⚙ ACTIVITY

The computer simulation in **Student Activity 2.15** looks at nucleic acid structure and the sequence of events in protein synthesis.

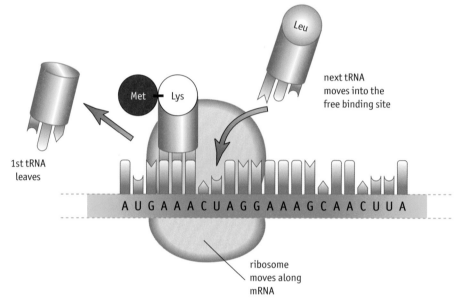

Figure 2.48 The ribosome moves along the mRNA, allowing another complementary tRNA to bind.

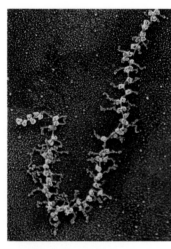

Figure 2.49 False-colour transmission electron micrograph showing translation. New proteins (green) are formed from the mRNA (pink). As the protein molecule is formed, it folds up into the three-dimensional shape determined by its primary structure – the sequence of amino acids (see page 63).

If several ribosomes attach to a single mRNA molecule (see Figures 2.41 and 2.49), several copies of the same protein can be produced at the same time.

Q 2.25 On which type of RNA would you find **a** a codon **b** an anticodon?

Q 2.26 The sequence of bases AGT form a triplet code on the sense strand. What is:

(a) its triplet code on the antisense strand

(b) its codon

(c) its anticodon.

Q 2.27 What would be the sequence of bases on a length of messenger RNA built using the following DNA strand as a template?

T A C A T G G A T T C C G A T

Q 2.28 How many tRNA molecules would be involved in the synthesis of the protein coded for by this section of DNA?

Q 2.29 What are the anticodons, assuming you read the section from left to right?

Q 2.30 Use Table 2.2 to determine which amino acids this sequence of DNA would code for, reading from left to right.

Q 2.31 Suggest why a change to a DNA base in a gene might not affect the primary amino acid sequence of a protein?

2.5 What can go wrong with DNA?

A mistake in transcription can produce mRNA with one or more incorrect codons. This could result in the production of a faulty protein or no protein at all, but because the fault is in the mRNA, it would only affect the proteins produced from this one mRNA strand in this one cell, on this one occasion. It would not produce the problems seen in every epithelial cell of cystic fibrosis sufferers. It is errors in the DNA that are responsible for inherited genetic conditions. These mistakes arise when DNA is copied during the process of DNA replication.

DNA replication

When a cell divides during the growth or repair of an organism, an exact copy of the DNA must be produced so that each of the new daughter cells receives a copy. This process of copying the DNA is called **replication**. Although the stages of replication show some similarities with those of transcription, there are fundamental differences and it important to realise they are quite distinct processes.

The entire DNA double helix unwinds from one end and the two single strands split apart as the hydrogen bonds between the bases break (Figure 2.50). Free DNA nucleotides line up alongside each single DNA strand and hydrogen bonds form between the complementary bases. The enzyme **DNA polymerase** links the adjacent nucleotides with phosphodiester bonds in condensation reactions to form new complementary strands. In this way each original strand of DNA acts as a template on which a new strand is built and, overall, two complete DNA molecules are formed. These are identical to each other and to the original DNA molecule. Check this by comparing them in Figure 2.50. Each of the two DNA molecules now contains one 'old' strand and one 'new' strand. This process is known as **semi-conservative replication**.

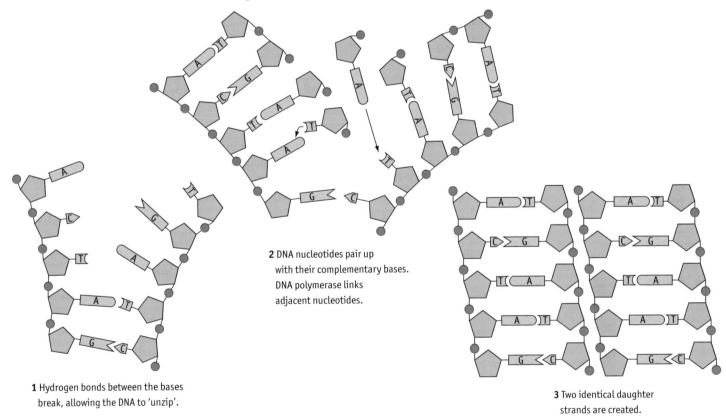

2 DNA nucleotides pair up with their complementary bases. DNA polymerase links adjacent nucleotides.

1 Hydrogen bonds between the bases break, allowing the DNA to 'unzip'.

3 Two identical daughter strands are created.

Figure 2.50 DNA replication. The DNA polymerase is not shown in this diagram, it catalyses the condensation reaction between adjacent nucleotides.

How do we know that DNA replication is semi-conservative?

In their letter to the science journal *Nature* in 1953 suggesting a double helical model of DNA, Watson and Crick also propose the specific pairing of bases A and T and bases C and G, with the sequence of bases on one chain automatically determining the sequence on the other chain. They note that this immediately suggests a mechanism for copying genetic material.

However, it took four more years and a series of elegant experiments by the American geneticists Matthew Meselson and Franklin Stahl to confirm experimentally what Watson and Crick suspected about how this copying mechanism happens.

There are three possible ways DNA *could* replicate – in a fragmentary way, semi-conservatively or conservatively (Figure 2.51).

Figure 2.51 uses colour in the different DNA strands to distinguish between 'old' and 'new' DNA and follow their progress. Meselson and Stahl used heavy and light strands of DNA. They did this by using DNA from *Escherichia coli* bacteria that had been grown in a medium containing only the heavy isotope of nitrogen, ^{15}N. All the nucleotides in the bacteria at the start of their experiment contained heavy nitrogen, and this made the DNA denser than normal (yellow in Figure 2.51).

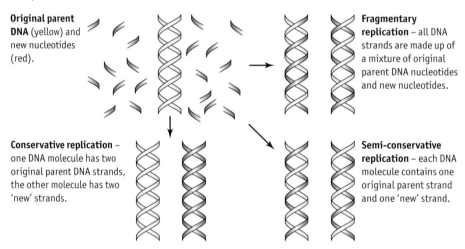

Original parent DNA (yellow) and new nucleotides (red).

Fragmentary replication – all DNA strands are made up of a mixture of original parent DNA nucleotides and new nucleotides.

Conservative replication – one DNA molecule has two original parent DNA strands, the other molecule has two 'new' strands.

Semi-conservative replication – each DNA molecule contains one original parent strand and one 'new' strand.

Figure 2.51 Three possible models of DNA replication.

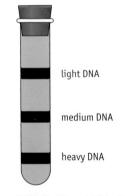

light DNA

medium DNA

heavy DNA

Figure 2.52 Position of DNA after density-gradient centrifugation.

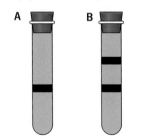

Figure 2.53 A Position of DNA after one replication. **B** Position of DNA after replicating twice.

Meselson and Stahl then moved the bacteria into a medium containing only normal ^{14}N. This meant all the new nucleotides (red in Figure 2.51) incorporated into the replicated DNA were 'light' but the original DNA nucleotides were 'heavy'. They allowed the bacteria to divide and their DNA to replicate once. They then extracted and centrifuged the DNA.

If a test tube containing DNA dissolved in a special density-gradient solution is centrifuged, the heavy DNA (containing only ^{15}N) sinks to the bottom. Light DNA (containing only normal ^{14}N) collects in a band near the top, and DNA of medium density is in the middle (Figure 2.52). Medium density DNA must contain some heavy and some light nucleotides.

Q 2.32 Where would you expect the bands of DNA to be after one round of replication according to each of the theories described in Figure 2.51?

Meselson and Stahl's result was a single band of medium density DNA (Figure 2.53A). Heavy DNA no longer existed, so they were able to reject the conservative model of DNA replication. However, which of the other two models was correct, as both would yield medium DNA after one round of replication? The answer lay in letting the bacteria undergo two rounds of replication.

Q 2.33 Using two different colours to represent the heavy and light bands, draw the DNA molecules after a second round of **a** fragmentary replication **b** semi-conservative replication.

The DNA extracted and centrifuged after two rounds of replication gave two bands – one medium and one light (Figure 2.53B). The presence of both medium and light DNA bands confirmed the semi-conservative model and ruled out the fragmentary model which would only produce one type of DNA containing a mixture of heavy and light nucleotides.

Q 2.34 Where would the DNA be in the test tube if the fragmentary model was correct?

Mistakes in replication

So what has all this to do with cystic fibrosis and other genetic disorders? Sometimes DNA replication does not work perfectly. As the 'new' strand of DNA is being built, inaccuracies with complementary base pairing sometime occur and an incorrect base may slip into place. This is an example of a gene mutation.

Q 2.35 Assuming no mutations, what sequence of bases would be created on the complementary strand of DNA by replication of an 'old' strand with the sequence

C A G T C A G G C?

Q 2.36 Identify the mutations that have occurred in replication of the same sequence in each of the following:

(a) G T C A G G C C G

(b) G A C A G T C C G

(c) G T C A T G C C G

(d) G T C G T C C G

(e) G T C A G G T C C G

Sometimes mutations occur in the DNA of an ovary or testis cell that is dividing to form an egg or sperm. Such a mutation may be passed on to future generations, present in every single cell produced from the fertilised egg.

Some mutations have no effect on the organism. Large amounts of the DNA found in a cell do not actually play a role in protein synthesis and therefore mutations that occur in these sections may have no effect. However, if a mutation occurs within a gene and a new base triplet is created that codes for a stop signal or a different amino acid, the protein formed may be faulty. This could cause a genetic disorder.

A mutation causes sickle cell anaemia

In the disease **sickle cell anaemia** there is a substitution mutation in the gene that codes for one of the polypeptide chains in haemoglobin, the pigment in red blood cells that carries oxygen around the body. The base adenine replaces thymine at one position along the chain. The mRNA produced from this DNA contains the triplet code GUA rather than GAA. As a result the protein produced contains the non-polar amino acid valine rather than polar glutamic acid at this point. This small change has a devastating effect on the functioning of the molecule. The haemoglobin is less soluble. When oxygen levels are low, the molecules form long fibres that stick together inside the red blood cell, distorting its shape. The resulting half moon (sickle) shaped cells carry less oxygen and can block blood vessels (Figure 2.54).

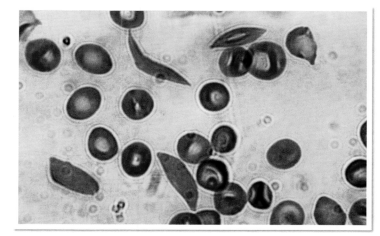

Figure 2.54 Can you identify the sickle-shaped red blood cells, which are symptoms of sickle cell anaemia? Magnification ×1200.

Mutations and cystic fibrosis

The CF gene is a section of DNA on chromosome 7 carrying the code to make the CFTR protein. The CFTR protein contains 1480 amino acids and these are arranged into the 3D structure shown in Figure 2.55.

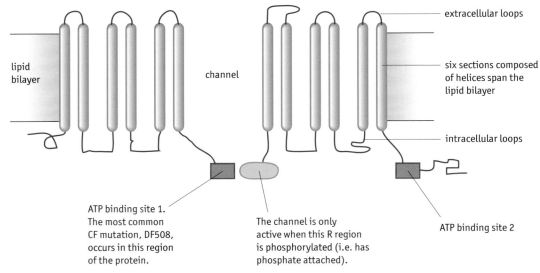

Figure 2.55 The CFTR protein is a channel protein.

Cystic fibrosis is not as straightforward a genetic story as sickle cell anaemia. In the CF gene hundreds of different mutations have been identified that can give rise to cystic fibrosis. The mutations affect the CFTR protein in different ways. In some cases ATP is unable to bind and open the ion channel; in other cases the channel is open but changes in the protein structure lead to reduced movement of chloride ions through the channel. The most common mutation, known as the DF508 mutation, is the deletion of three nucleotides. This causes the loss of phenylalanine, the 508th amino acid in the CFTR protein, which is thought to result in misfolding of the protein. The mutations are passed from parent to offspring.

Do Claire and Nathan have one of these mutations? Will they pass it on and if they do, will the child have the disease?

2.6 How is cystic fibrosis inherited?

Claire and Nathan know that cystic fibrosis is inherited but they need to understand *how* it is inherited. This will tell them the chances of their children inheriting the disease.

How genes are passed on

Genes and chromosome pairs

You know that a gene is a length of DNA that codes for a protein. Every cell (except the sex cells) contains *two* copies of each gene, one from each parent. For any particular gene the two copies are located in the same position, or **locus**, on each of two paired chromosomes. Humans have 23 pairs of chromosomes in our cells, and the chromosomes in each pair are called **homologous chromosomes**. Within each of these pairs, one chromosome originally came from our mother and the other from our father. Different species have different numbers of homologous pairs but all individuals of the same species have the same number of pairs.

Cystic fibrosis is caused by a gene mutation in the CF gene, carrying the code to make the CFTR protein. The gene is located on chromosome 7 and at the same locus on its homologous chromosome. The mutation is passed on from parents to their children. But consider the following three situations:

1 A couple has six children. The first five are healthy; the sixth has cystic fibrosis. Neither parent has the disease.

2 Another couple has two children; both children have cystic fibrosis. Again, neither parent has the disease.

3 A woman with cystic fibrosis is told that it is unlikely that her children will have the disease, but they will all be 'carriers' and could themselves have children with the illness.

So what is going on?

Genotypes, phenotypes and alleles

As we have seen, cystic fibrosis is caused by a mutation in the CF gene, the length of DNA which codes for the CFTR protein. The CF gene occurs in two alternative forms or **alleles**. First, there is the normal allele which codes for the functioning CFTR protein; this can be represented by the letter **F**. Secondly, there is the mutated allele which produces a non-functional protein; this can be represented by the letter **f**. Since every human has two copies of this gene in all their body cells there are three possible combinations that can occur, namely:

1 **FF** – A person with two identical copies of the normal allele does not have cystic fibrosis.

2 **ff** – A person with two copies of the mutated allele has cystic fibrosis.

3 **Ff** – A person with one normal allele and one mutated allele does not have cystic fibrosis but is a **carrier** and could have children who have the disease.

The alleles that a person has make up their **genotype**. Combinations 1 and 2, **FF** and **ff**, show a **homozygous** genotype for the CF gene – there are two identical copies of the allele. Combination 3 has two different alleles, **Ff**, and is **heterozygous** for the CF gene (Figure 2.56).

The characteristic (observable effect) caused by the genotype, is the **phenotype**. Table 2.3 summarises the cystic fibrosis genotypes and phenotypes.

F is called the **dominant allele**. It affects the phenotypes of one of the homozygotes (a person who has the homozygous genotype, **FF**) and the heterozygote (a person with the heterozygous genotype, **Ff**). On the other hand, **f** is the **recessive allele**; it only affects the phenotype of the other homozygote (**ff**).

If two CF carriers had children what genotypes and phenotypes would we expect their children to have?

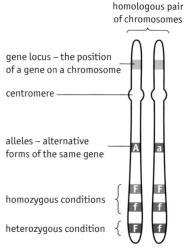

homologous pair of chromosomes

gene locus – the position of a gene on a chromosome

centromere

alleles – alternative forms of the same gene

homozygous conditions

heterozygous condition

The dominant allele is written as an upper case (capital) letter.

Figure 2.56 Some basic genetic vocabulary.

Genotype	Phenotype
FF	normal
ff	cystic fibrosis
Ff	normal, but carrier

Table 2.3 The relationship between genotype and phenotype for the cystic fibrosis gene.

Predicting the genotypes of offspring

In the UK about 1 person in 25 is a cystic fibrosis carrier (**Ff**). They can pass the disease on to their children. When gametes are produced, each egg or sperm contains only one allele, in this case either **F** or **f**. These two types of gamete are produced in roughly equal numbers. The expected genotypes of children produced by two cystic fibrosis carriers can be shown in a genetic diagram, known as a Punnett square. This illustrates all the possible ways in which the two types of allele can combine, and thus shows the possible genotypes that could occur in the children.

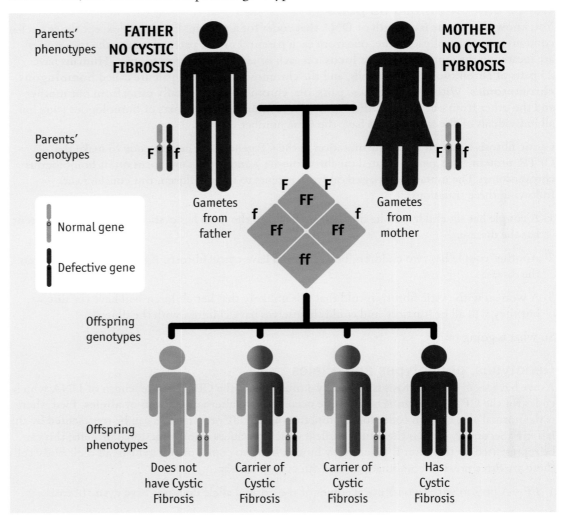

This means that every time a child is born to parents who are both carriers of cystic fibrosis, there is a 1 in 4 or 0.25 probability that the baby will have the genotype **ff** and have cystic fibrosis. There is also a 1 in 4 probability that the genotype will be **FF**, and a 1 in 2 (2 in 4) or 0.5 probability that it will be **Ff**, a CF carrier.

Until very recently people with the genotype **ff** did not often survive to be adults and therefore did not have any children. You might think that natural selection would have eliminated such a harmful allele, but in fact heterozygotes (**Ff**) have some protection against the dangerous disease typhoid. So in areas where typhoid was common, carriers would have been at a definite advantage. Today, people with CF are much more likely to reach adulthood but most have reduced fertility because of the sticky mucus in the reproductive system. IVF can help these individuals have a baby.

ACTIVITY
You can use Reebops to investigate inheritance in **Student Activity 2.17**.

Q 2.37 Give a genetic explanation for each of the three situations described at the top of page 93.

Q 2.38 The genetic pedigree diagram below shows the inheritance of CF for three generations of one family. Explain how Frank could have inherited CF from James and Margaret who do not have CF.

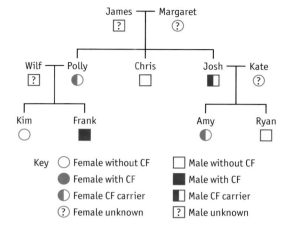

Key
○ Female without CF   □ Male without CF
● Female with CF   ■ Male with CF
◐ Female CF carrier   ◧ Male CF carrier
⍰ Female unknown   ☐? Male unknown

Inheritance of some human genetic diseases

Cystic fibrosis is an example of **monohybrid inheritance**, so called because the characteristic is controlled by only one gene. Most human characteristics are inherited in a much more complex way, and are often influenced by environmental factors. However, there are a few characteristics controlled by single genes.

For example, **thalassaemia** is another genetic disease affecting the production of haemoglobin. The condition is caused by recessive alleles of a gene on chromosome 11. A number of different mutations can affect this gene. Someone who is homozygous for one of these recessive alleles either makes no haemoglobin at all, or makes haemoglobin that cannot carry out its function. The homozygous condition is often eventually lethal.

People who are heterozygous show no symptoms, but have some protection against malaria. For this reason thalassaemia is relatively common in people who live (or have ancestors who lived) in areas where malaria occurs or occurred in the past, particularly around the Mediterranean Sea. Because of this 'heterozygous advantage' the mutant alleles have not disappeared.

Other conditions such as **albinism** and **phenylketonuria** are also caused by single recessive alleles. **Achondroplasia** (a form of dwarfism), on the other hand, is caused by a dominant allele. A homozygote, carrying two copies of the allele for achondroplasia, always dies. Someone who is heterozygous for this condition will show very restricted growth. **Huntington's disease** and the ability to taste PTC (a bitter-tasting chemical called phenylthiocarbamide) are also caused by dominant alleles.

You should not suppose that characteristics determined by a single gene are found only in humans. They occur in all organisms. Indeed, they were first discovered in the garden pea by Gregor Mendel in the 1850s and 1860s. There are also characteristics that show incomplete dominance where neither allele is dominant and heterozygotes have an intermediate phenotype. For example, in snapdragons, a plant homozygous for red flowers crossed with a plant homozygous for white flowers produces heterozygous offspring with pink flowers because one allele is not dominant.

The work of Mendel

Mendel is rightly known as the father of genetics. He was a monk and carried out a huge number of breeding experiments in the garden of his Moravian monastery. Mendel died without other scientists appreciating the significance of his work. However, Mendel established that a number of characteristics of the garden pea were determined by separate 'heredity units' or what he called factors and we now call genes. For example, whether the pea plants are tall (1.9–2.2 m in height) or short (0.3–0.5 m) is determined by one gene. Whether the seeds are smooth or wrinkled is determined by another gene.

WEBLINK
Investigate the relationship that also exists between sickle cell anaemia and resistance to malaria in the case studies on the DNA to Darwin website.

ACTIVITY
Student Activity 2.18 lets you apply ideas about inheritance to some other situations.

CHECKPOINT
2.6 Produce a vocabulary list giving definitions of the following key genetic terms:
- gene
- locus
- homologous chromosomes
- alleles
- dominant allele
- recessive allele
- heterozygous
- homozygous
- carrier
- genotype
- phenotype.

2.7 Testing for CF

Even though 1 person in 25 in the UK is a CF carrier, the first time that most carriers realise that they carry the CF allele is when one of their children is born with the disease.

Once cystic fibrosis is suspected, it is possible to carry out conventional tests to confirm the diagnosis. Ancient European folklore warned that a child who tasted salty when kissed would die young, and this has become the basis of a modern test that measures the level of salt in sweat. The test works because CF-affected people have markedly higher concentrations of salt in their sweat.

People with CF also have elevated levels of the protein trypsinogen in their blood. Testing for this protein is part of the Newborn Blood Spot Screening Programme for all babies in the UK. The early diagnosis of CF allows treatment to begin immediately, which can improve health in later years.

Genetic testing

The gene that codes for the CFTR protein was identified and sequenced in 1989. This led to the possibility of **genetic testing**, identifying the abnormal allele of the gene in the DNA of any cells. This in turn has paved the way for **genetic screening** to confirm the results of conventional tests, to identify carriers and also to diagnose CF in an embryo or fetus. The terms 'genetic testing' and 'genetic screening' are sometimes distinguished but often, as here, used interchangeably.

Genetic testing can be performed on any DNA. Samples of cheek cells or white blood cells can be taken from a parent and the DNA tested to find out whether they are a carrier of the disease. Cells obtained from a fetus or embryo can be tested to establish whether it will be affected by the disease. The DNA is screened to see whether it contains the known base sequences for the most common mutations that cause cystic fibrosis.

 Q 2.39 The test is performed only for the most common mutations that cause CF. What does this mean about the reliability of the test? Will there be any false positives or negatives?

How can genetic screening be used?

To confirm a diagnosis

Genetic testing can confirm an initial diagnosis of CF, which might have been based on sweat testing or the newborn blood spot trypsinogen test. However, since there are a large number of different mutations of the *CFTR* gene which cause the disease, a negative result must be treated with caution. It is not currently feasible to test for all of the hundreds of possible mutations that lead to CF.

To identify carriers

Genetic testing can identify carriers. A sample of blood or cells taken from inside the mouth can be used to detect abnormal alleles in people without the disease who are heterozygous. Where there has been a history of cystic fibrosis in a family this can be of value in assessing the probability of having a child with the disease. Counselling is offered before and after testing, and parents can make informed decisions about how to proceed.

For testing embryos

Currently there are two well-established and widely-used techniques for prenatal genetic testing (the testing of DNA from an embryo or fetus) and a third technique in the early stages of development. Note that an embryo becomes a fetus at 10 weeks of pregnancy, or 8 weeks after conception. Pregnancy is counted from the first day of the woman's period before she conceives. The most common method of fetal testing is **amniocentesis**, which involves inserting a needle into the amniotic fluid to collect fetal cells that have fallen off the placenta and fetus (Figure 2.57). When amniocentesis is carried out, usually at around 15–17 weeks of pregnancy, there is about a 1% risk of causing a miscarriage.

Chorionic villus sampling (CVS) is the second invasive technique. Here, a small sample of placental tissue (which includes cells of the embryo or fetus) is removed, either through the wall of the abdomen or through the vagina. CVS can be carried out earlier in pregnancy, between 8 and 12 weeks, since there is no need to wait for amniotic fluid to develop. However, it carries a slightly higher risk of inducing a miscarriage than 15 week amniocentesis. The Royal College of Obstetricians and Gynaecologists 2010 guidelines on amniocentesis and CVS do not quantify the risk as there has been very limited research undertaken. The NHS website quotes an estimated risk of about 1% to 2% of inducing a miscarriage but provides no evidence to support this claim.

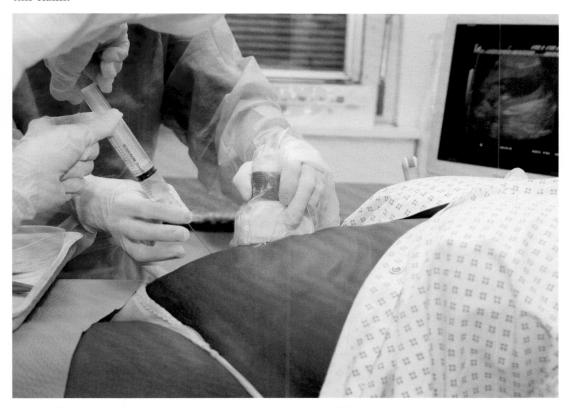

Figure 2.57 A prenatal test during pregnancy such as amniocentesis or CVS involves techniques that carry some risk.

There are implications associated with the use of these tests which people should be aware of when deciding to have the tests so they can make informed choices. Both amniocentesis and CVS procedures present a risk of miscarriage to possibly healthy fetuses. If any of the tests give a positive result for the disease, one option is for the woman to have an abortion. Having an abortion is easier for the woman, both physically and emotionally, in the earlier weeks of pregnancy. This can mean that the slightly higher risk of miscarriage associated with CVS is considered worthwhile.

Q 2.40 Are there circumstances where people might choose not to test, even if the test was offered to them?

A third technique, **non-invasive prenatal diagnosis** (NIPD), is now offered to some couples. NIPD works by analysing DNA fragments in the mother's blood plasma during pregnancy. Whilst most of this 'cell-free DNA' is from the mother herself, about 10–20% is from the embryo. Cell-free fetal DNA (cffDNA) becomes detectable in the mother at about 4–5 weeks of pregnancy; however, at this stage the levels are too low to be analysed. Samples are likely to be collected after seven or nine weeks of pregnancy, depending on the genetic test to be performed as different genetic tests require differing concentrations of cffDNA before a test can be carried out.

WEBLINK
To find out more about NIPD visit the RAPID (Reliable Accurate Prenatal non-Invasive Diagnosis) project website.

The method has been used in screening for a limited number of single gene disorders. For example, if the mother does not have achondroplasia but it is thought that the fetus may have inherited the condition, the cell-free DNA from the mother's blood sample is tested for the dominant allele that results in achondroplasia. If found, it can be concluded that the embryo has inherited the condition from the father. He may have the condition or may have passed on a new gene mutation that occurred during formation of the sperm. The latter accounts for 80–90% of cases and is associated with older fathers.

NIPD has also been used, on a research basis, where a couple already know they are both CF carriers and their mutations are different. Cell-free DNA is isolated from the mother's blood and screened. If the father's CF mutation is not found in the cell-free DNA, the baby cannot have CF, although it may be a carrier. NIPD has also been used for the diagnosis of chromosomal conditions such as Down's Syndrome. Because NIPD only requires a blood sample from the mother there is no risk of miscarriage as a result of an invasive testing procedure. NIPD is a rapidly developing technology and it is likely that it will be available for a wider range of genetic conditions in the future.

Testing before implantation (PGD)

If a couple have a family history of a serious genetic condition like cystic fibrosis or already have a child with the condition, pre-implantation genetic diagnosis (PGD) may be an option. The couple will have to undergo *in vitro* fertilisation (IVF) in order to create embryos that can be tested before transfer to the uterus. When an early embryo is growing in culture and has around eight cells, one cell can be removed for genetic testing without harming the embryo (Figure 2.58). The DNA of the cell is analysed and the results of the genetic screening are used to decide whether to place the embryo into the uterus. IVF is an expensive and sometimes stressful procedure and although it avoids the need for a possible abortion, success rates are quite low. The live birth rate for PGD is similar to general IVF success rates, at around 30% for women under 35. In 2010, only 121 babies were born in the UK as a result of PGD.

Making ethical decisions – what is right and what is wrong?

How should we decide in life what is right and what is wrong? For example, should we always tell the truth? Can we ever justify turning down a request for help? Should Claire and Nathan have some form of genetic testing? Should Claire have an abortion if they find that their unborn baby has cystic fibrosis?

All of us have **moral** views about these and other matters. For example, you might feel that lying and abortion are always wrong and helping people always right. But in order to maintain that something is ethically acceptable or unacceptable, you must be able to provide a reasonable explanation as to *why* that is the case.

There is no one universally accepted way of deciding whether something is ethically acceptable or not. What there are instead are a number of ethical frameworks, each of which allows you to work out whether a particular action would be right or wrong if you accept the ethical principles on which the framework is based. Usually you get the same answer whichever framework you adopt. But not always! This is why perfectly thoughtful, kind and intelligent people sometimes still disagree completely about whether a particular course of action is justified or not.

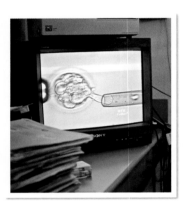

Figure 2.58 Why can one cell be removed at this stage without harming the developing embryo? We will see the answer to this question in Topic 3.

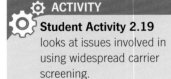

ACTIVITY

Student Activity 2.19 looks at issues involved in using widespread carrier screening.

Ethical frameworks

We shall examine four widely-used ethical frameworks. You should find these of value when considering various issues, such as genetic screening and abortion, raised in this topic. We will also refer to these frameworks in other topics in this course.

1 Rights and duties

Most of us tend to feel that there are certain human **rights** that should always be permitted. For example, we talk about the right to life, the right to a fair trial and the right to freedom of speech. Certain countries, for example the USA, have some of these rights enshrined in their constitutions.

If you have a right to something, then I may have particular **duties** towards you. For example, suppose that you are a six-month-old baby with a right to life and I am your parent. I have a duty to feed you, wash you, keep you warm and so on. If I do not fulfil these duties, I am failing to carry out my responsibilities and the police or social services may intervene.

But where do rights come from? Some people with a religious faith find them in the teachings of their religion. For example, the ten commandments in the Jewish scriptures talk about not stealing, not murdering, telling the truth and so on.

But nowadays many people, indeed in the UK most people, have little or no religious faith. So where can they – perhaps you – find rights? The simplest answer is that rights are social conventions built up over thousands of years. If you want to live in a society you have, more or less, to abide by its conventions.

2 Maximising the amount of good in the world

Perhaps the simplest ethical framework says that each of us should do whatever *maximises the amount of good in the world*. For example, should I tell the truth? Usually yes, as telling lies often ends up making people unhappy and unhappiness is not good. But sometimes telling the truth can lead to more unhappiness. If your friend asks you if you like the present they have just given you and you do not, would you tell the truth? Most of us would tell a 'white lie', not wanting to hurt their feelings.

This ethical approach is known as **utilitarianism**. Notice that utilitarians have no moral absolutes beyond maximising the amount of good in the world. A utilitarian would hesitate to state that anything is always right or always wrong. There might be circumstances in which something normally right (e.g. keeping a promise) would be wrong and there might be circumstances when something normally wrong (e.g. killing someone) would be right.

3 Making decisions for yourself

One of the key things about being a human is that we can make our own decisions. There was, for example, a time when doctors simply told their patients what was best for them. Now, though, there has been a strong move towards enabling patients to act autonomously. People act autonomously when they make up their own minds about things. If you have ever had an operation you will probably have signed a consent form. The thinking behind this is that a surgeon should not be allowed to operate on you unless you have given **informed consent**.

Of course, it is perfectly possible autonomously to decide to be absolutely selfish! A utilitarian would say we need to weigh the benefits of someone acting autonomously with any costs of them doing so. Only if the overall benefits are greater than the overall costs is **autonomy** desirable. Someone who believes in rights and duties might say that each of us has a right to act autonomously but also has a duty to take account of the effects of our actions on others.

Not everyone can make decisions for themselves. For example, people with learning difficulties or dementia may not be able to give informed consent. In these situations an advocate may be appointed to help them make a decision or to make the decision on their behalf.

> **CHECKPOINT**
>
> **2.7** Use each of these four ethical frameworks to consider whether or not it is acceptable to abort a fetus found by amniocentesis to have CF.

4 Leading a virtuous life

A final approach is one of the oldest. This holds that the good life (in every sense of the term) consists of acting virtuously. This may sound rather old fashioned but consider the **virtues** that you might wish a good teacher/lecturer to have. She or he might be understanding, be able to get you to learn what you want or need to learn, and believe in treating students fairly.

Traditionally the seven virtues were said to be **justice**, prudence (i.e. wisdom), temperance (i.e. acting in moderation), fortitude (i.e. courage), faith, hope and charity. Precisely what leading a virtuous life means can vary and is not always straightforward. Think about the virtues you might like to see in a parent, a doctor and a girl-/boyfriend. What would be the virtuous course of action for Claire and Nathan?

Genetic counselling

If a couple like Nathan and Claire are at risk of having a child with a genetic disorder, a genetic counsellor can provide advice. Such a person will help the couple understand how the disease is inherited and the chance that any child they conceive will have it. A genetic counsellor will explain the tests available and the possible courses of action, depending on the outcome of the tests. The counselling should help the couple decide such things as whether to be tested themselves, whether to have children, use *in vitro* fertilisation with screening or use prenatal screening.

Q 2.41 If both parents are found to be carriers, what are the options open to them?

To help make their decision Claire and Nathan also need to know what treatment options are available for CF if they are to make informed decisions they need to know what will be involved if they accepted the risks and their child did inherit CF.

DID YOU KNOW?

Treatments for CF

1 Medication to relieve symptoms

There are a wide range of medications commonly used to relieve the symptoms of CF. These include the following.

- Bronchodilators – these are drugs that are inhaled using a nebuliser. Air is blown through a solution to make a fine mist which is breathed into the lungs through a mouthpiece. The drugs relax the muscles in the airways, opening them up and relieving tightness of the chest.

- Antibiotics – the early diagnosis and treatment of lung infection is the cornerstone of CF treatment. Many antibiotics are used to kill or prevent the growth of bacteria in the lungs.

- DNAase enzymes – infection of the lungs leads to the accumulation of white blood cells in the mucus. The breakdown of these white blood cells releases DNA, which can add to the 'stickiness' of the mucus. DNAase enzymes can be inhaled using a nebuliser. They break down the DNA, so the mucus is easier to clear from the lungs.

- Steroids are used to reduce inflammation of the lungs.

Medication to treat CF

Ivacaftor is the first of a new class of drug called a CFTR potentiator. It is only effective for CF sufferers who have at least one copy of the *CFTR* mutation G551D. This mutation means the amino acid glycine at position 551 in the CFTR protein is replaced by aspartate so ATP no longer binds, which prevents the ion channel from opening. Exactly how ivacaftor works is unclear but it is thought to bind somewhere near the mutation and open the ion channel. Clinical trials of ivacaftor, taken as a pill twice a day, have shown that it improves lung function, lowers sweat chloride levels and helps patients gain weight. The NHS policy is that ivacaftor should be routinely prescribed for patients over the age of 6 with at least one copy of G551D – around 320 people in the UK. The annual cost of treatment with ivacaftor is in the region of £180 000 per patient per year (2014 cost).

2 Diet

Adults with CF are recommended to eat high-energy foods and their diet should include double the quantity of protein recommended for people who do not have CF. Some people with CF may also need salt supplements.

3 Digestive enzyme supplements

If the pancreatic duct is blocked, the food molecules in the small intestine cannot be broken down far enough to be absorbed. Taking enzyme supplements with food helps to complete the process of digestion.

4 Physiotherapy

Rhythmical tapping of the walls of the chest cavity (percussion therapy) and use of a flutter device can help loosen the mucus and improve the flow of air into and out of the lungs (Figure 2.59). Such treatment needs to be carried out regularly, twice a day.

Figure 2.59 Children learn how to perform their own physiotherapy. Here a child uses a flutter device. Exhaling through the flutter's special valve causes rapid changes in air pressure within the airways. The vibrations aid movement of the mucus.

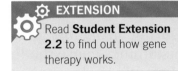
EXTENSION
Read **Student Extension 2.2** to find out how gene therapy works.

5 Heart and lung transplant

If the lungs become badly damaged and very inefficient, other treatments may become ineffective at relieving the symptoms. The only option available may be to replace the damaged lungs with a heart and lung transplant.

6 Possible CF treatments for the future – gene therapy

Gene therapy alters the genotype and hence the phenotype of cells affected by the condition. Normal alleles of the gene are inserted into the target cells, using a vector (for example, a virus) or a carrier mechanism (for example, liposomes). The normal form of the gene is transcribed and translated, to produce a functioning protein. In CF gene therapy trails a working CFTR protein is produced and incorporated into the cell membrane, thus restoring the ion channel and avoiding the symptoms of CF.

Trials of CF gene therapy started in the UK back in 1993 and the 'normal' CFTR allele has been successfully transferred to the lung epithelial cells of CF patients. In early trials, chloride transport in the lungs was restored to 25% of normal. However, this type of improvement was temporary. Cells are continuously lost from the epithelium lining the airways so the transfer of the allele to these cells does not offer a permanent solution. The longest the correction has lasted in any trial is about 15 days. Although CF gene therapy research has not yet resulted in significant therapeutic success, gene therapy has been more successful with other conditions, such as severe combined immunodeficiency (SCID). It is hoped that in future gene therapy will help in the treatment of CF. The UK's first major multi-dose therapeutic trial CF was due to finish in 2014, the results were not known at the time of publication.

Most of the current treatments alleviate the symptoms of CF, unfortunately at the moment none provide a cure. Effective gene therapy would treat the cause by altering the affected cells' genotypes so functioning CFTR proteins are produced in the epithelial cells. This treatment is still at the trail stage. Claire and Nathan need to bear this in mind when making their decision about starting a family.

ACTIVITY
The role play in **Student Activity 2.20** lets you think about some of the issues covered in this topic.
In **Student Activity 2.21** read a personal experience of another gene mutation to help you revisit some ideas considered earlier in the topic.

ACTIVITY
Use **Student Activity 2.22** to check your notes using the topic summary provided.

ASTHMA

A student researching asthma viewed many online resources including the two sources below.

ASTHMA

Asthma causes the airways of the lungs to narrow so people have difficulty in breathing. People with asthma have over-sensitive airways that become irritated by triggers such as pollen, house dust mites, pet hairs, exercise, smoke or even cold air. Asthma can also be triggered by stress. Someone who has asthma isn't affected all the time. They may have attacks several times a day or only a few times a year.

During an asthma attack the cells lining the bronchioles release chemicals called **histamines**. The histamines cause the lining cells to become inflamed, produce large amounts of mucus and swell. Histamines also make the muscles in the walls of the bronchioles contract. As a result of all these changes the airways narrow, making it very difficult to move air into and out of the lungs.

Extract from an on-line resource on breathing and asthma produced by the Association of the British Pharmaceutical Industry (ABPI) to support biology education in UK schools. Aimed at GCSE students, it pushes the levels of knowledge on towards the beginning of A level studies.

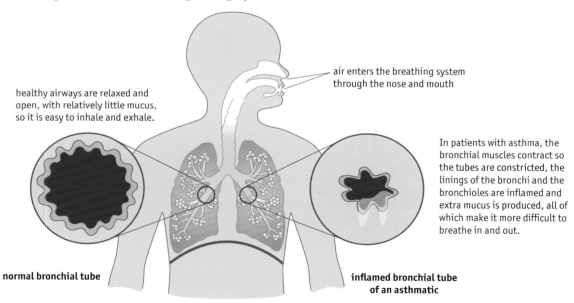

healthy airways are relaxed and open, with relatively little mucus, so it is easy to inhale and exhale.

air enters the breathing system through the nose and mouth

In patients with asthma, the bronchial muscles contract so the tubes are constricted, the linings of the bronchi and the bronchioles are inflamed and extra mucus is produced, all of which make it more difficult to breathe in and out.

normal bronchial tube

inflamed bronchial tube of an asthmatic

The changes in the bronchioles during an asthma attack can make it difficult to breathe

Links across the course

1.1   1.2   1.3   1.4   2.1   2.2   2.3   2.4   2.5   2.6

Published in the Daily Express: 10:54, Wed, July 2, 2014   By Jo Willey

CHILDHOOD OBESITY IS FUELLING ASTHMA EPIDEMIC, NEW RESEARCH WARNS

Obesity is a growing problem among children in the UK

Experts have found that the risk of developing the potentially-fatal breathing condition increases by 55 per cent for every extra unit of Body Mass Index (BMI).

Young children are getting fatter with nearly 30 per cent aged two to 15 are either overweight or obese – one of the highest rates in Western Europe.

Previous research has suggested that expanding waistlines could be causing more cases of childhood asthma.

But in a new study, experts at Bristol University have directly linked obesity to an increased risk of the respiratory disease.

The incidence of asthma, a chronic condition caused by inflammation of the airways, has been rising steadily over the past few decades with over 1.1 million children having asthma in the UK.

Latest figures suggest 18.9 per cent of children in Year 6 aged 10–11 were obese and a further 14.4 per cent were overweight.

Of the children in Reception aged 4–5, latest figures suggest 9.3 per cent were obese and another 13.0 per cent were overweight. In total almost a third of 10–11 year olds and over a fifth of 4–5 year olds were overweight or obese.

Although the underlying causes of asthma are not known some experts think obesity may be one of the causes.

The new study published in PLOS Medicine used both genetic information and observational data in order to assess whether BMI can actually cause asthma.

The study looked at 4,835 children with asthma by the age of seven-and-a-half enrolled in the Avon Longitudinal Study of Parents and Children.

A weighted genetic score based on 32 independent BMI-related DNA sequence variations was calculated, and associations with BMI, fat mass, lean mass, and asthma were estimated.

The research found that the genetic score was strongly associated with BMI, fat mass, and lean mass, and with childhood asthma, and that the relative risk of asthma increased by 55 per cent for every extra unit of BMI.

These findings suggest that a higher BMI increases the risk of asthma in mid-childhood, and that public health interventions designed to reduce obesity may also help to limit the global rise in asthma.

However the researchers noted it is possible that the observed association between BMI and asthma could have been affected by some of the genetic variants included in the BMI genetic

score which could also independently increase the risk of asthma.

There was also some evidence that body composition – lean body mass – influences asthma through pathways not related to obesity-induced inflammation.

Dr Raquel Granell said: "Environmental influences on the development of asthma in childhood have been extensively investigated in epidemiological studies, but few of these provide strong evidence for causality."

"Higher BMI in mid-childhood could help explain some of the increase in asthma risk toward the end of the 20th century, although the continued rise in obesity but with a slowing in the rise in asthma prevalence in some countries implies that other non-BMI-related factors are also likely to be important."

START BY REVIEWING THE SOURCE

Read the extracts and think critically about them as sources of scientific information.

1. Comment critically on the reliability of each extract as a source of scientific information, support your views with appropriate evidence.

2. If you were recommending some additional sources for the student to use what features would you say she should look for to be confident that they are a reliable source of scientific information?

REVIEWING THE BIOLOGY

Having read the article, draw on your knowledge gained so far in the course and answer the following questions.

1. Explain how an asthma attack will affect efficiency of gas exchange.

2. Compare and contrast the causes and effects of asthma and cystic fibrosis on lungs.

3. 1 in 11 children and 1 in 12 adults in the UK are affected by asthma. They take either relieving or preventative medication with more people taking relieving medication. Suggest how you think these two types of medications treat asthma.

4. The newspaper article links BMI and asthma, discuss whether you would consider this to be a correlation or causal link, support your discussion with evidence from the article.

5. Suggest what the article might mean by BMI-related DNA sequence variations.

6. Comment on the quality of the epidemiological research study reported in the article above.

At the end of your A level course come back and answer the question below.

1. What additional information would it be valuable for a doctor to have when determining the effect of an asthma attack on a patient's breathing and gas exchange?

Information sources
As you read the articles, consider where they have come from, who wrote them and whom they were written for. Established scientific publications are good sources of reliable information, whereas other resources might be less dependable for a number of reasons. Think about what makes a source reliable and why.

VOICE OF THE GENOME

Why a topic called Voice of the genome?

Most of the multicellular organisms that can be seen around us, from apple trees to zebras, including magnolias, mice, monkeys, mushrooms, mayflies and us, *Homo sapiens*, start out in much the same way – as a very simple, undifferentiated cell. But from that cell come scores of different cells, specialised for a huge variety of jobs and all produced in exactly the correct places. How is the amazing change from a single cell into a complex body achieved (Figure 3.1)? In this topic, we follow the fate of the gametes produced by parents as these gametes start on the greatest journey of all – from separate cells to a full-grown adult.

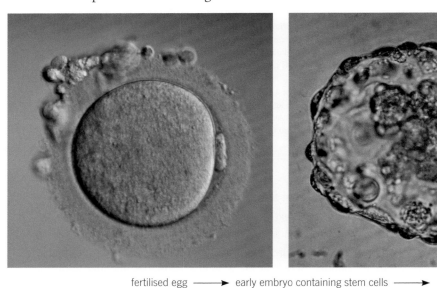

fertilised egg ⟶ early embryo containing stem cells ⟶

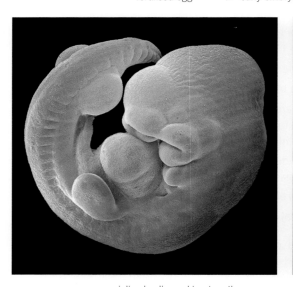

specialised cells working together ⟶ complex multicellular organism ⟶

Figure 3.1 From a single cell to a full-grown adult.

William Harvey (1578–1657), one of the greatest anatomists who ever lived, said 'Everything comes from eggs'. But people used to think that the whole organism was wrapped up in miniature inside the sperm, ready to grow and be unfolded during development (Figure 3.2). Of course this is not the case, but the truth is not so different. A fertilised egg contains a complete set of instructions. Inside every fertilised egg and body cell is a complete genome – DNA containing a full set of genes that control the growth and development of the whole organism.

Figure 3.2 It was once thought that a miniature baby was to be found inside the sperm cell.

In humans, there are approximately 20 000 to 25 000 genes. We share 98.5% of our DNA sequence with chimpanzees, making them our closest relatives in the animal kingdom. However, we also share 50% of our genes with bananas, which is food for thought. Genes provide the instruction manual to build and control not only a single cell, but also the great range of specialised cells that make up an adult organism. How do these genes talk to the different cells?

In some organisms, features you might think are controlled exclusively by genes are not quite so straightforward, with genes and the environment interacting. What is the relationship between genes and the environment? Is cancer inherited, caused by our environment, or a combination of both? What is the role of the epigenome?

OVERVIEW OF THE BIOLOGICAL PRINCIPLES COVERED IN THIS TOPIC

Gamete structure and function provide the starting point for this topic. You will look at their role in fertilisation. To understand how a single cell divides and grows into a whole organism, you need a good grasp of cell structure and a detailed knowledge of the cell cycle. At GCSE, the basics of cell division were covered. Here, you will gain a more detailed picture of this process, including the role of both nuclear and cytoplasmic division.

You will discover how the first cells produced have the potential to develop into any part of the body and see how these stem cells are becoming increasingly important to medical research. They offer a huge potential for the creation of new treatments and therapies. But these new developments are controversial. You will have to consider the arguments for and against the use of stem cells in medicine, in order to clarify your own views on the issue.

You will learn how cells become specialised through differential gene expression and how increasing knowledge about epigenetics is changing our understanding about the control of gene expression. You will also find out how individual specialised cells work together as tissues, organs and organ systems.

You will also look at how the characteristics of an organism are controlled by both genetic make-up and the environment.

REVIEW

Are you ready to tackle Topic 3 *Voice of the genome*?

Complete the GCSE review and GCSE review test before you start.

3.1 In the beginning

Have you ever tried incubating an egg bought from the supermarket (Figure 3.3)? If so, it is unlikely that the egg hatched because these eggs are unfertilised. The hens they come from are kept in female-only flocks: no cockerels are present, so no sperm have combined with these eggs. To get a chick, you need a fertilised egg that divides and differentiates into a complex multicellular body.

The life cycle of most multicellular organisms starts in this way. Gametes combine to form a single fertilised egg that develops into a new adult made up of numerous specialised cells organised into tissues and organs. But are all of the cells basically the same as that fertilised egg?

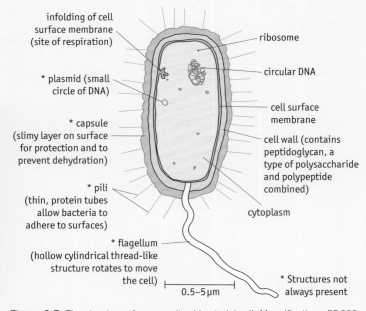

Figure 3.3 Which came first?

KEY BIOLOGICAL PRINCIPLE: TWO DIFFERENT TYPES OF CELLS

All living organisms are made of cells.

Cells have many common features, but it is possible to distinguish two basic types: **prokaryotic** and **eukaryotic** cells.

Prokaryotic cells

Bacteria and cyanobacteria (photosynthetic bacteria) together make up the Prokaryotae kingdom (see Topic 4). Their cells do not have nuclei or other membrane-bound cell organelles (Figures 3.4 and 3.5). This type of cell is called a 'prokaryotic cell', meaning 'before the nucleus'.

Most prokaryotes are extremely small with diameters between 0.5 and 5 μm. Their DNA is not associated with any proteins and lies free in the cytoplasm. A cell wall is always present in prokaryotic cells.

infolding of cell surface membrane (site of respiration)

ribosome

* plasmid (small circle of DNA)

circular DNA

* capsule (slimy layer on surface for protection and to prevent dehydration)

cell surface membrane

cell wall (contains peptidoglycan, a type of polysaccharide and polypeptide combined)

* pili (thin, protein tubes allow bacteria to adhere to surfaces)

cytoplasm

* flagellum (hollow cylindrical thread-like structure rotates to move the cell)

0.5–5 μm

* Structures not always present

Figure 3.5 The structure of a generalised bacterial cell. Magnification ×55 000.

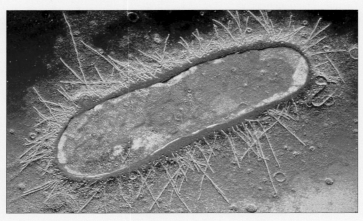

Figure 3.4 Electron microscope image of the bacterium *Vibrio cholerae*. Magnification ×55 000.

Eukaryotic cells

All other living organisms have cells that contain discrete membrane-bound organelles, such as nuclei, mitochondria and chloroplasts (plants only). These are 'eukaryotic cells', meaning 'true nucleus'. Eukaryotic cells are larger than prokaryotic cells, with diameters of 20 μm or more. Unlike the prokaryotes, not all eukaryotic cells have a cell wall. Organisms with eukaryotic cells are eukaryotes.

Inside the eukaryotic animal cell

The classical child's diagram of a cell resembles a fried egg: one circle with a large blob in the middle. Using a simple light microscope, this is almost as much detail as it is possible to see. However, the use of electron microscopy reveals a wealth of structures within the cell, often described as the cell ultrastructure (Figure 3.6).

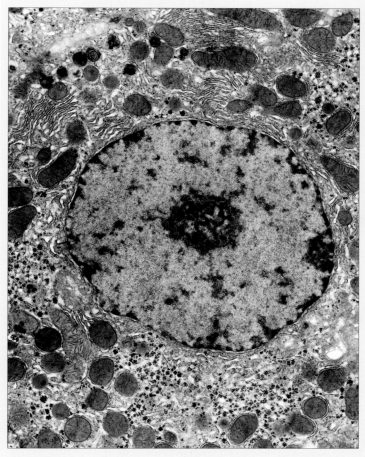

Figure 3.6 Electron micrograph of a liver cell showing the cell's ultrastructure. The nucleolus is clearly visible at the centre of the nucleus. The dark oval and round structures are mitochondria, and the black specks are glycogen granules. Magnification ×10 200.

Figure 3.7 shows a two-dimensional representation of the cross-section of a generalised animal cell. You can see that there are many structures within the cell cytoplasm. These are shown in three dimensions in Figure 3.8. Use these diagrams to answer the questions below.

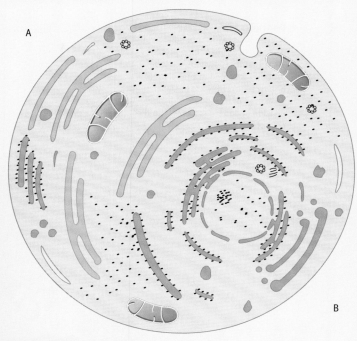

Figure 3.7 A generalised animal cell.

Q 3.1 **(a)** Follow a straight line across the cell in Figure 3.7 from A to B and identify each of the structures that the line crosses using Figure 3.8.

(b) Identify two structures within the cell that are not crossed by this line.

Q 3.2 You can work out the actual size of an electron micrograph image by measuring it in μm (mm × 1000) and dividing this by the magnification. Work out the diameter of the nucleus in Figure 3.6.

The structures and functions of plant cells will be studied in Topic 4.

⚙ **ACTIVITY**

In **Student Activity 3.1** you look in more detail at the relationship between the three-dimensional structures and the functions of organelles in the cell, using the interactive cell.

Mitochondrion (plural mitochondria) The inner of its two membranes is folded to form finger-like projections called cristae. The mitochondria are the site of the later stages of aerobic respiration.

Centrioles Every animal cell has one pair of centrioles, which are hollow cylinders made up of a ring of nine protein microtubules (polymers of globular proteins arranged in a helix to form a hollow tube). They are involved in the formation of the spindle during nuclear division and in transport within the cell cytoplasm.

Nucleus Enclosed by an envelope composed of two membranes perforated by pores. Contains chromosomes and a nucleolus. Chromosomes made of DNA contain genes that control the synthesis of proteins.

Nucleolus A dense body within the nucleus where ribosomes are made.

Rough endoplasmic reticulum (rER) A system of interconnected membrane-bound, flattened sacs. Ribosomes are attached to the outer surface. Proteins made on these ribosomes are transported through the ER to other parts of the cell.

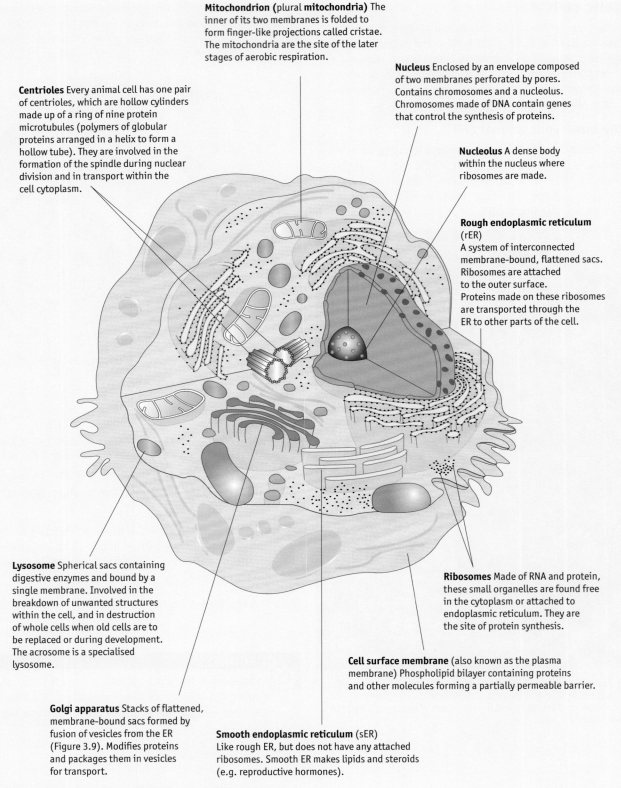

Lysosome Spherical sacs containing digestive enzymes and bound by a single membrane. Involved in the breakdown of unwanted structures within the cell, and in destruction of whole cells when old cells are to be replaced or during development. The acrosome is a specialised lysosome.

Ribosomes Made of RNA and protein, these small organelles are found free in the cytoplasm or attached to endoplasmic reticulum. They are the site of protein synthesis.

Cell surface membrane (also known as the plasma membrane) Phospholipid bilayer containing proteins and other molecules forming a partially permeable barrier.

Golgi apparatus Stacks of flattened, membrane-bound sacs formed by fusion of vesicles from the ER (Figure 3.9). Modifies proteins and packages them in vesicles for transport.

Smooth endoplasmic reticulum (sER) Like rough ER, but does not have any attached ribosomes. Smooth ER makes lipids and steroids (e.g. reproductive hormones).

Figure 3.8 A three-dimensional representation of a generalised animal cell.

Q 3.3 Look at the electron micrograph in Figure 3.9 and identify the structures labelled **A** to **E**.

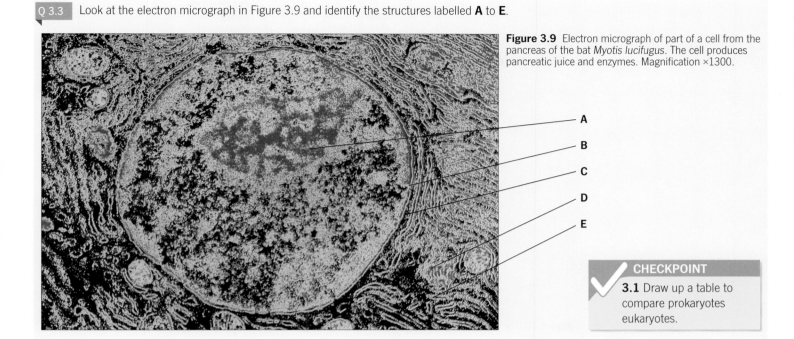

Figure 3.9 Electron micrograph of part of a cell from the pancreas of the bat *Myotis lucifugus*. The cell produces pancreatic juice and enzymes. Magnification ×1300.

A
B
C
D
E

> **CHECKPOINT**
>
> **3.1** Draw up a table to compare prokaryotes eukaryotes.

DID YOU KNOW?

The origins of chloroplasts and mitochondria

Chloroplasts and mitochondria are thought to have originally been independent prokaryotes. Very early in the evolution of multicellular organisms, a mutualistic relationship (of benefit to both partners) developed with eukaryotic cells. Perhaps these ancestors of chloroplasts and mitochondria gained protection and a more constant environment by being inside eukaryotic cells. The eukaryotes probably benefited from the products of the metabolic processes of the prokaryotes: sugar and oxygen from the photosynthetic prokaryotes that became chloroplasts; ATP from the non-photosynthetic prokaryotes that became mitochondria. To this day, both chloroplasts and mitochondria contain their own DNA, separate from the DNA of the cells they inhabit. However, they can no longer replicate themselves independently and rely partly on enzymes made by their hosts.

Cells are dynamic

The cell is not static: there is continuous movement of molecules within it. In Topic 2, you saw the movement of RNA out of the nucleus into the cytoplasm, where protein synthesis takes place. Once the proteins have been synthesised on ribosomes, they are processed and move through the cell to where they are needed. In addition, many proteins, such as enzymes, hormones and signal proteins, are released from the cells where they are made. This movement of proteins through the cell involves the endoplasmic reticulum, Golgi apparatus and vesicles. These membrane structures are continuously created and lost within the cell, as outlined in Figure 3.10.

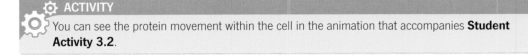

ACTIVITY

You can see the protein movement within the cell in the animation that accompanies **Student Activity 3.2**.

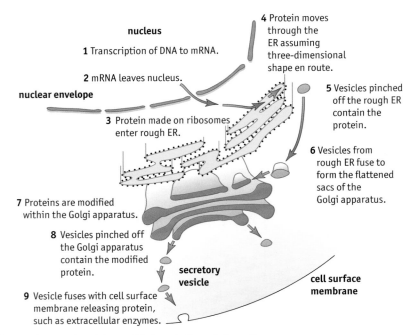

Figure 3.10 The production of proteins and their route through the cell.

Getting started

All eukaryotic cells share the same basic structures and many functions. As multicellular organisms develop, cells take on specialised roles. Gametes, the starting point in the story of development, are highly specialised, with structures and functions that are obviously different from other body cells.

Gametes

Mammalian gametes

You will remember from GCSE that in mammals the **sperm** and **ovum** (egg) cells are the **gametes** or sex cells (Figure 3.11).

The gametes are adapted for their roles in sexual reproduction. In humans, as in other mammals, the ovum is a large cell incapable of independent movement. It is wafted along one of the oviducts from the ovary to the uterus (Figure 2.35A, page 79) by ciliated cells lining the tubes and by muscular contractions of the tubes. The cytoplasm of the ovum contains protein and lipid food reserves for a developing embryo. Surrounding the cell is a jelly-like coating called the zona pellucida.

The sperm cell is much smaller than the ovum and is motile (can move). To enable it to swim, the sperm cell has a long tail (flagellum), powered by energy released by mitochondria. Males continuously produce large numbers of sperm once they have reached maturity. Sperm that enter the vagina during intercourse swim through the uterus, their passage assisted by muscular contractions of the uterus walls. If intercourse takes place at about the time of ovulation, sperm may meet the ovum in the oviduct. The sperm are attracted to the ovum by chemicals released from it. The **acrosome** in the head of the sperm swells, fuses with the sperm cell surface membrane and releases digestive enzymes. These break down the zona pellucida of the ovum (Figure 3.12). This is called the **acrosome reaction**. The acrosome is a type of lysosome. Lysomes are enzyme-filled sacs found in the cytoplasm of many cells.

Once a sperm fuses with and penetrates the membrane surrounding the egg, chemicals released by the ovum cause the zona pellucida, to thicken, preventing any further sperm entering the egg. This process is called the **cortical reaction**. The sperm nucleus that enters the egg fuses with the egg nucleus to produce a fertilised egg (zygote).

CHECKPOINT

3.2 Look at Figure 3.11. Make a list of the differences between the mammalian gamete cells, and explain the value of each feature listed.

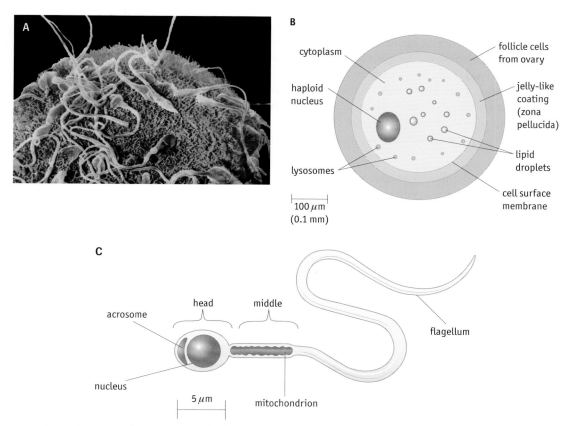

Figure 3.11 **A** Scanning electron micrograph of sperm on an ovum during fertilisation. Magnification ×1600. **B** Diagram of a human ovum. **C** Diagram of a human sperm.

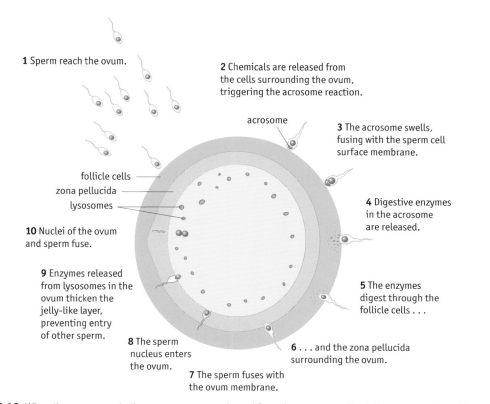

1 Sperm reach the ovum.

2 Chemicals are released from the cells surrounding the ovum, triggering the acrosome reaction.

acrosome

3 The acrosome swells, fusing with the sperm cell surface membrane.

follicle cells
zona pellucida
lysosomes

4 Digestive enzymes in the acrosome are released.

10 Nuclei of the ovum and sperm fuse.

9 Enzymes released from lysosomes in the ovum thicken the jelly-like layer, preventing entry of other sperm.

5 The enzymes digest through the follicle cells . . .

8 The sperm nucleus enters the ovum.

6 . . . and the zona pellucida surrounding the ovum.

7 The sperm fuses with the ovum membrane.

Figure 3.12 When the sperm meets the ovum, enzymes released from the acrosome digest the outer coating of the ovum.

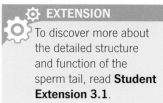

ACTIVITY

In **Student Activity 3.3** you can relate the structures of mammalian gametes to their functions.

You can watch fertilisation in marine worms in **Student Activity 3.4**.

EXTENSION

To discover more about the detailed structure and function of the sperm tail, read **Student Extension 3.1**.

Gamete cells are unusual

Although the number of chromosomes differs from species to species, all 'normal' individuals within a species will have the same number of chromosomes in each cell. Human cells contain 46 chromosomes made up of 22 homologous pairs and one pair of sex chromosomes; fruit fly cells contain eight chromosomes made up of three homologous pairs and one pair of sex chromosomes; white campion plant cells have 24 chromosomes made up of 11 homologous pairs and one pair of sex chromosomes; rice cells contain 24 chromosomes made up of 12 homologous pairs (there are no sex chromosomes as separate male and female individuals are not produced); and so on.

The fundamental difference between gametes and other cells is the number of chromosomes they contain. In Topic 2, you saw that gametes have half the number of chromosomes found in all other body cells – one chromosome from each homologous pair – but have you considered why this is? Think about what happens at fertilisation. If the sperm and ovum cells both had the full chromosome number, which is 46 in humans, then the zygote would have 92 chromosomes. When this individual reproduced the zygote would contain 184 chromosomes. With each generation the number of chromosomes would double, which obviously would not work. This is why the gametes are haploid containing half the full number of chromosomes – 23 in humans – made up of one of each homologous pair and one sex chromosome. When the gametes fuse, the full number of 46 is restored as you can see in Figure 3.13. A diploid zygote is formed.

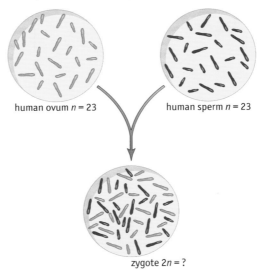

human ovum $n = 23$

human sperm $n = 23$

zygote $2n = ?$

Figure 3.13 Haploid gametes come together to form a diploid zygote. What number should replace the question mark? Count them if you are not sure. The number of chromosomes in gametes and zygotes varies from species to species. Human zygotes contain 46 chromosomes, dogs have 78 and peas have 14.

How do gametes form?

There are two different types of cell division in living organisms. One involves **mitosis** and produces new body cells as an organism grows and develops. This retains the full number of chromosomes, called the **diploid** number ($2n$) (46 in humans). The other type of cell division involves **meiosis** and produces gametes with only half the number of chromosomes, called the **haploid** number (n) (23 in humans). This form of cell division occurs in the ovaries and testes of animals, and the ovaries and anthers of flowering plants. This type of cell division is shown in Figure 3.14.

Meiosis has two important roles. Firstly, it results in haploid cells, which are necessary to maintain the diploid number after fertilisation. Secondly, it helps create genetic variation among offspring.

How does meiosis result in genetic variation?

The shuffling of existing genetic material into new combinations during meiosis is important in creating genetic variation. This shuffling includes: **independent assortment** and **crossing over**.

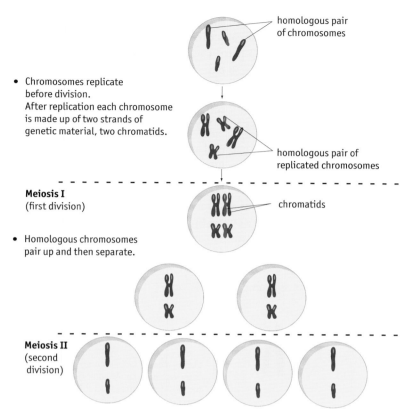

- Chromosomes replicate before division.
  After replication each chromosome is made up of two strands of genetic material, two chromatids.

Meiosis I (first division)

- Homologous chromosomes pair up and then separate.

Meiosis II (second division)

- Chromatids separate and gametes are formed, each with half the original number of chromosomes.

Figure 3.14 Gamete production by meiosis.

Independent assortment

During meiosis only one chromosome from each pair ends up in each gamete. The independent assortment of the chromosome pairs as they line up during meiosis is a source of genetic variation. This process is random: either chromosome from each pair could be in any gamete. This way of sharing out chromosomes produces genetically variable gametes. The way that this happens is shown in Figure 3.15. An organism with six chromosomes, that is three homologous pairs **XX**, **YY** and **ZZ**, could form eight (2^3) combinations in its gametes.

The arrangement of each chromosome pair during the first division in meiosis is completely random. In a cell with three chromosome pairs both the arrangements shown here are possible, for example.

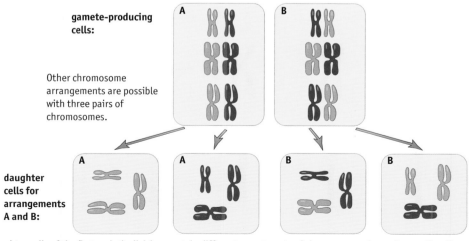

gamete-producing cells:

Other chromosome arrangements are possible with three pairs of chromosomes.

daughter cells for arrangements A and B:

The daughter cells of the first meiotic division contain different assortments of chromosomes depending on the alignment of the chromosomes during the first meiotic division.

Figure 3.15 Independent assortment introduces variation.

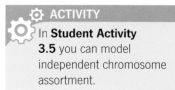

ACTIVITY

In **Student Activity 3.5** you can model independent chromosome assortment.

When these join with another set at fertilisation, this pretty well guarantees that individuals produced from sexual reproduction are genetically different from each other. In organisms with many chromosomes, such as humans with 23 pairs, the number of possible combinations of the chromosomes is so large that it is unlikely any two daughter cells will have the same chromosome combination.

Q 3.4 Give the possible combinations of chromosomes that could occur in the gametes of an organism with three homologous pairs **XX**, **YY** and **ZZ**.

Q 3.5 How many possible combinations of maternal and paternal chromosomes could be found in the gametes of organisms with $2n = 8$; and organisms with $2n = 10$?

Crossing over

During the first meiotic division, homologous chromosomes come together as pairs and all four **chromatids** come into contact. At these contact points the chromatids break and rejoin, exchanging sections of DNA between non-sister chromatids (see Figure 3.16). The point where the chromatids break is called a **chiasma** (plural **chiasmata**) and several of these often occur along the length of each pair of chromosomes, giving rise to a large amount of variation. There is no crossing over between the sex chromosomes during meiosis.

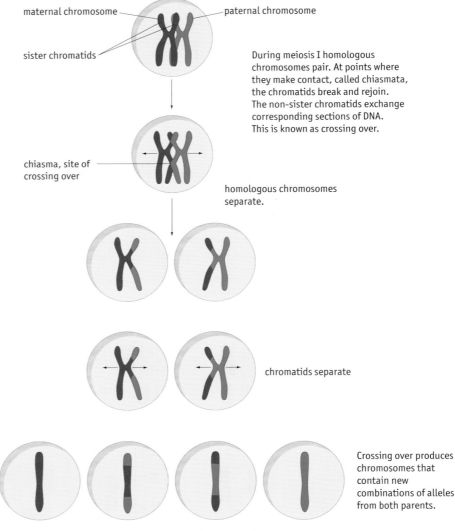

maternal chromosome — paternal chromosome

sister chromatids

During meiosis I homologous chromosomes pair. At points where they make contact, called chiasmata, the chromatids break and rejoin. The non-sister chromatids exchange corresponding sections of DNA. This is known as crossing over.

chiasma, site of crossing over

homologous chromosomes separate.

chromatids separate

Crossing over produces chromosomes that contain new combinations of alleles from both parents.

Figure 3.16 Crossing over can result in a great deal of genetic variation.

Q 3.6 What would the chromosomes in the gametes produced from the homologous chromosomes in cells A and B look like after crossing over has occurred?

A

Linkage

We have seen that independent assortment leads to new combinations of chromosomes and results in new combinations of genes in the gametes, producing genetic variation. But is the inheritance of one gene independent of other genes? Do all genes become mixed equally during meiosis?

In the 1850s, Mendel analysed the patterns of inheritance of characteristics of garden peas. He crossed pure breeding, homozygous, tall purple-flowered plants with short white-flowered plants. All of the offspring produced were tall with purple flowers. He went on to cross two of these plants and found that plants were produced showing all combinations of the characteristics: tall and purple, tall and white, short and purple, and short and white. The outcome of the cross is shown below.

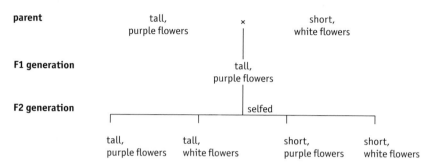

From these results, Mendel concluded that the inheritance of one pair of factors (he did not know about genes at the time) is independent of the inheritance of other pairs. However, it has since been found that Mendel's idea is not always the case: there are some characteristics that tend to be inherited together. In these instances, one particular allele for a gene is inherited with one particular allele for another gene – they appear to be linked together. We now know that genes are inherited independently only if they are on separate chromosomes or are far apart on the same chromosome. Any two genes with a locus on the same chromosome are linked together and will tend to be passed as a pair to the same gamete. This is known as **linkage** of genes. Figure 3.17 shows the effect of linkage of two genes in fruit flies (*Drosophilia* spp.). The genes will only be separated and go into

B

CHECKPOINT

✓ **3.3** Produce a concept map or table that summarises how genetic variation is generated.

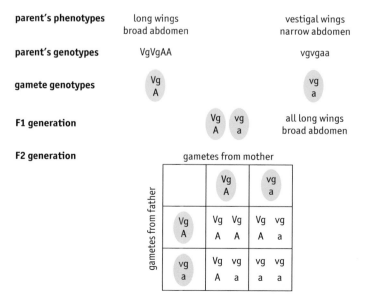

Figure 3.17 When homozygous fruit flies (*Drosophilia* spp.) are mated, the first generation (F1) all have long wings and broad abdomens; none have vestigial (short misshapen) wings and narrow abdomens. When these heterozygous individuals are crossed they do not produce the full range of characteristics expected if the genes were inherited independently. The genes are linked, so the offspring have either long wings and broad abdomens or vestigial wings and narrow abdomens, with a 3:1 ratio typical of monohybrid inheritance.

⚙ **ACTIVITY**

In **Student Activity 3.6** mate some 'virtual' fruit flies to discover how genes can be linked.

different gametes if crossing over happens between the pair of genes. If the two genes are very close together on the same chromosome, crossing over is very unlikely to happen between them. The two genes are said to be strongly linked and they will be inherited as a pair.

Q 3.7 **(a)** If the genes in peas for tall and purple flowers were linked and always inherited together, and so were the short and white flowers genes, what would be the outcome of mating two tall purple plants?

(b) Look at Figure 3.17. If the two characteristics were not linked, what combination of features would result from a cross between two heterozygous individuals?

Genes on a single chromosome make up a linkage group. In humans, there are 23 linkage groups because we have 23 pairs of chromosomes. We can picture each of the 23 chromosomes as a chain of genes, with the longer chromosomes having more genes in the chain. Chromosomes are numbered from the longest to the shortest, so we expect chromosome 1 to contain the most genes and chromosome 22 the fewest. Chromosome 23 is the sex chromosome.

Q 3.8 Explain why there are four linkage groups in fruit flies (*Drosophilia* spp.).

It is now known on which chromosome and at which locus on the chromosome most of our genes are situated. For example, the *CFTR* gene responsible for cystic fibrosis is located from base pair 117 470 771 to base pair 117 668 664 on chromosome 7.

Sex linkage

As you probably know, in mammals, including humans, sex is controlled by the **sex chromosomes**. Females are **XX** and males are **XY**. The Y chromosome contains genes that make a person male and very few other genes. All the genes on the sex chromosomes are passed on with those that determine sex, they are **sex-linked**.

Many people have imperfect vision and sometimes this is in the form of colour blindness. Colour blindness means not being able to see all colours clearly. The most common type is the confusion of red and green colours. Rather surprisingly, you are much more likely to be red-green colour blind if you are male than if you are female.

Red-green colour blindness occurs in about 8% of men, but in only about 0.5% of women. Some other phenotypes, including the blood disease, haemophilia, are also much more common in men than in women. These conditions are called sex-linked. In fruit flies, eye colour is sex linked, with males having white eyes rather than the usual red eyes, more often than females.

You can check whether or not you have any form of colour blindness using an Ishihara test. These can be viewed online. If you have any questions or worries about your colour vision you should consult your doctor or optician.

Red-green colour blindness

The gene loci for the three pigments found in the eye's cone cells are on the X chromosome. Cone cell pigments are needed for colour vision. If one or more cone pigments is faulty or absent, the person will not be able to see colours normally: they will have some form of colour blindness.

How is colour blindness inherited?

Red-green colour blindness happens when there is a mutation in one of the cone pigment genes. This could happen (very rarely) when an egg or sperm is produced in meiosis.

Let us start with a woman called May who is heterozygous for colour blindness, $X^N X^n$. A mutation has occurred in the gene for cone pigment in one of the gametes that formed her.

X^N is an X chromosome with a normal allele N for cone pigment.

X^n is an X chromosome with a mutant allele n for cone pigment.

May has normal vision because allele N is dominant to allele n: she has one allele that works normally and so can make cone pigment. She is a carrier of colour blindness.

May has children with George. George is $X^N Y$, so he also has normal colour vision. Note that the Y chromosome has no allele symbol (N or n) because the Y chromosome has no locus for this gene.

	May	George
Parents' phenotypes	Full colour vision	Full colour vision
Parents' genotypes	$X^N X^n$	$X^N Y$

Gamete genotypes

		Gametes from mother	
		X^N	X^n
Gametes from father	X^N	$X^N X^N$	$X^N X^n$
	Y	$X^N Y$	$X^n Y$

Offsprings' phenotypes: 50% Full colour vision girl (25% are carriers), 25% Full colour vision boy, 25% Colour blind boy

The genetic diagram shows that there is a 25% probability that the offspring from these two parents will be colour blind, but only colour blind boys will occur and not colour blind girls.

Q 3.9 Explain how it is possible for a colour blind girl to occur.

Q 3.10 A couple had children using IVF because the father has haemophilia, a sex-linked condition. Only female embryos were used. Explain whether or not future generations of this family would be free of haemophilia.

Getting together

Fertilisation in mammals

To produce a new individual, the nuclei from the gametes combine in the process of **fertilisation** (Figure 3.12). In mammals, the nucleus from one sperm enters the ovum, and the genetic material of the ovum and sperm fuse, forming a fertilised ovum called a **zygote**. This cell now contains genetic material from both parents.

For many organisms, the diploid cell formed at fertilisation is just the starting point on the road to creating a more complex multicellular structure. This single cell divides and gives rise to numerous specialised cells, creating the huge variety of structures within an organism.

DID YOU KNOW?

One cause of male infertility

For the human zygote to develop, the gamete nuclei have to fuse and a chemical from the sperm cytoplasm is required to activate the fertilised cell. This chemical is a protein called oscillin. It causes calcium ions to move in and out of stores in the cytoplasm of the ovum. These oscillations of calcium ion concentration trigger the zygote to begin developing into an embryo. Oscillin is concentrated in the first part of the sperm to attach to the ovum and enters before the male nucleus, activating the ovum. It is thought that low levels of oscillin in sperm may be linked to male infertility and this is a current area of research.

EXTENSION

To find out about sperm competition read **Student Extension 3.2**.

Gametes and fertilisation in plants

Plant gametes

In flowering plants, gametes are produced by meiosis. The female gametes, ova, are produced within the ovary, but only one of the four nuclei at the end of meiosis forms an ovum — the others disintegrate. Inside the anther, cells divide to produce pollen grains, which contain the male gamete nuclei. In male plants, the four haploid cells produced at the end of meiosis undergo another division, this time by mitosis, where each nucleus divides to give two haploid nuclei within the pollen grain.

Fertilisation in flowering plants

In flowering plants, nuclei from the gametes also have to combine in the process of fertilisation (Figures 3.18). Fertilisation takes place within the ovule. The pollen grain germinates on the style and a pollen tube grows down through the style towards the ovary, with its growth controlled by one of the haploid nuclei from the pollen, the tube nucleus. On germination of the pollen, the other haploid nucleus, the generative nucleus, divides to form two haploid gamete nuclei that move down the pollen tube. The tube grows through a microscopic

pore into the embryo sac inside the ovule and the two male gamete nuclei enter the sac. One fuses with the egg cell and forms a diploid zygote, and the second fuses with two nuclei in the embryo sac, called polar nuclei, to form a triploid cell. The diploid zygote divides to form the embryo. The triploid cell divides to form the seed's storage tissue, the endosperm.

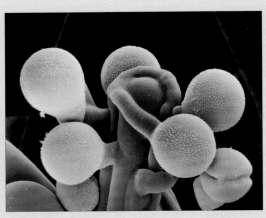

Figure 3.18A Scanning electron micrograph of germinating pollen grains. Pollen tubes are visible growing into the stigma. Magnification ×390.

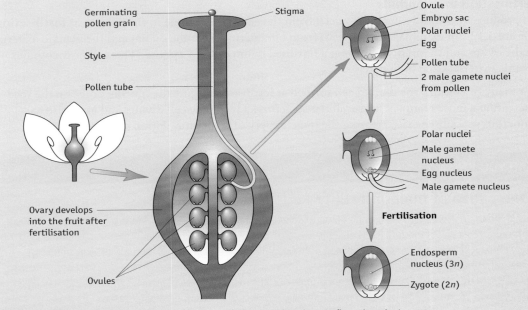

Figure 3.18B Double fertilisation involving two male gametes is unique to flowering plants.

3.2 From one to many: the cell cycle

Elephants (Figure 3.19) have much bigger bodies than mice. However, elephant cells are little or no larger than the corresponding mouse cells – there are just far more of them. Elephants, mice, trees and poppies all start as a single fertilised egg with the potential to divide and grow into the complete body of a new individual. To make the vast number of cells required to build individuals, new cell contents must be synthesised and then one cell must divide into two. This well-organised pattern of events is called the **cell cycle**.

Figure 3.19 How many cells are there in an elephant? Humans have approximately ten thousand thousand million (10^{13}) cells.

The cell cycle

The cell cycle can be divided into two distinct parts: interphase and division (Figure 3.20).

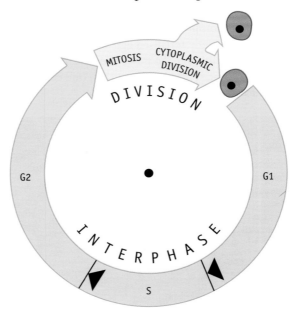

Figure 3.20 The stages in the cell cycle.

Preparation for division: interphase

Interphase is a time of intense and organised activity during which the cell synthesises new cell components, such as organelles and membranes, and new DNA. The formation of new cellular proteins occurs throughout interphase, whereas DNA synthesis occurs during the **S** (for synthesis of DNA) **phase**. The S phase separates the first gap or G_1 **phase** from the second gap or G_2 **phase**, as shown in Figure 3.20.

The length of interphase differs depending on the role of the cell. In the developing human embryo, there is no interphase for the first few divisions – the zygote already contains the materials needed to form the first 16 or so cells. In these first few divisions, the embryo divides without growing in size, producing smaller cells with each cell cycle. This makes the embryonic cell cycle much faster than those of other body cells.

The S and G_2 phases of most cells remain relatively constant in duration. The length of the G_1 phase is more variable: some cells can take weeks, months or even years to complete this phase. For example, liver cells may divide only once every one or two years due to an extended G_1 phase. In seeds, the cells of the embryo usually remain in the G_1 phase until germination occurs. Some cells, such as nerve and muscle cells, never divide again, remaining permanently in a non-dividing state. In plants, cell division is localised in small groups of cells called meristems, with most other cells not dividing.

Q 3.11 Suppose the cell cycle shown in Figure 3.20 lasts approximately 24 hours. Calculate the approximate length of time that the G_1, S, G_2 and division phases will last in this cell cycle.

Q 3.12 Look at Figure 3.21. What do you notice about the nucleus during interphase? What does this suggest to you about what may be going on inside the cell?

The interphase nucleus is a fairly uniform, featureless structure with one or two darker-staining regions called nucleoli (singular nucleolus). Ribosomes are formed in the nucleoli and they give nucleoli their dark appearance in electron micrographs. Ribosomes are made of protein and rRNA (ribosomal RNA). The rest of the nucleus contains the chromosomes, which are made up of DNA associated with proteins (Figure 3.22).

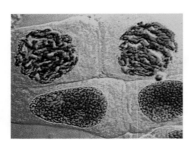

Figure 3.21 The lower two cells are typical of interphase.

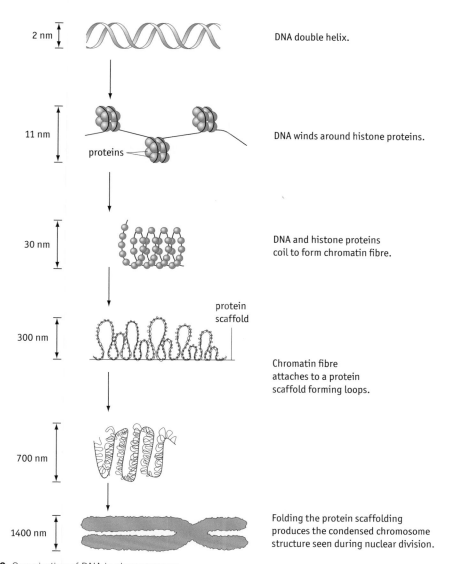

2 nm — DNA double helix.

11 nm — DNA winds around histone proteins.

proteins

30 nm — DNA and histone proteins coil to form chromatin fibre.

protein scaffold

300 nm — Chromatin fibre attaches to a protein scaffold forming loops.

700 nm

1400 nm — Folding the protein scaffolding produces the condensed chromosome structure seen during nuclear division.

Figure 3.22 Organisation of DNA in chromosomes.

During interphase, the individual chromosomes are unravelled (Figure 3.22). This allows access to the genetic material, enabling new proteins to be synthesised. In preparation for cell division, the cell synthesises additional cytoplasmic proteins and organelles. The cell must also produce copies of DNA for the two new cells. It is vital that this DNA is identical in both structure and quantity to the DNA in the original cell. This is achieved by DNA replication, as described in Topic 2, page 89.

Q 3.13   Which organelles would you expect to be most active in a cell that is producing large quantities of new proteins?

Cell division

By the end of interphase, the cell contains enough cytoplasm, organelles and DNA to form two new cells. The next step is to share out both the DNA and the contents of the cytoplasm so that each new cell can function independently. The DNA is separated in **nuclear division** (mitosis). **Cytoplasmic division** follows this.

Cell division is a continuous process, where a single cell with double the usual amount of cell contents becomes two new cells. However, it is possible to describe four stages during nuclear division by the behaviour of the chromosomes and other structures within the cell. Figure 3.23 shows what is happening in each of these stages of mitosis, known as **prophase**, **metaphase**, **anaphase** and **telophase**.

ACTIVITY

In **Student Activity 3.7** try to work out what happens in each of the four stages of mitosis using the mitosis flick book.

Compare this with the cell cycle/mitosis animation in **Student Activity 3.8**.

INTERPHASE

During **interphase**, new cell organelles are synthesised and DNA replication occurs. By the end of interphase, the cell contains enough cell contents to produce two new cells.

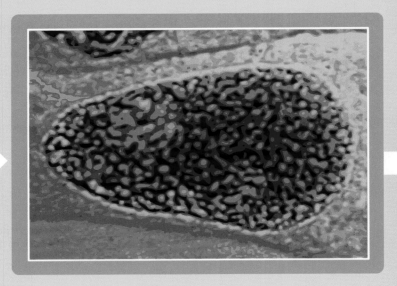

CYTOPLASMIC DIVISION

After nuclear division, the final reorganisation into two new cells occurs. This is called **cytoplasmic division**. In animal cells, the cell surface membrane constricts around the centre of the cell. A ring of protein filaments bound to the inside surface of the cell surface membrane is thought to contract until the cell is divided into two new cells. It has been proposed that the proteins actin and myosin, responsible for muscle contraction, may also be the proteins responsible for cytoplasmic division. Instead of undergoing this constriction, plant cells synthesise a new cell plate between the two new cells (Figure 3.24 A overleaf).

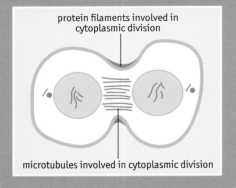

protein filaments involved in cytoplasmic division

microtubules involved in cytoplasmic division

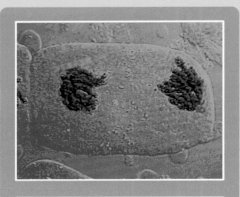

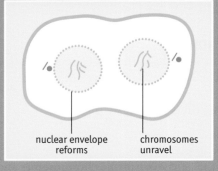

TELOPHASE

This last stage of mitotic division is called **telophase**. This is effectively the reverse of prophase. The chromosomes unravel and the nuclear envelope reforms, so that the two sets of genetic information become enclosed in separate nuclei.

nuclear envelope reforms

chromosomes unravel

Figure 3.23

Follow the sequence of events during cell division, clockwise in the direction of the arrows. The diagrams show a generalised animal cell containing only four chromosomes.

PROPHASE

During **prophase**, the chromosomes condense, becoming shorter and thicker, with each chromosome visible as two strands called **chromatids**. Apart from the occasional mutation, the two strands are identical copies of one another, produced by replication. They are effectively two chromosomes joined at one region called the **centromere** (Figure 3.25 overleaf).

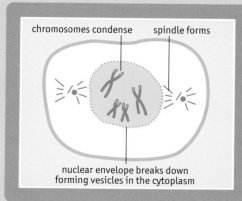

chromosomes condense   spindle forms

nuclear envelope breaks down forming vesicles in the cytoplasm

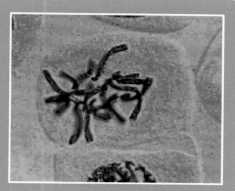

Q 3.14   Why do the chromosomes condense?

During prophase, microtubules from the cytoplasm form three-dimensional structure called the **spindle**. The **centrioles** move around the nuclear envelope and position themselves at opposite sides of the cell. These form the two poles of the spindle and are involved in the organization of the spindle fibres. The spindle fibres form between the poles. The widest part of the spindle is called the equator. The breakdown of the nuclear envelope signals the end of prophase and the start of the next stage.

Q 3.15   Why does the nuclear envelope disintegrate?

LATE ANAPHASE

In the next stage of mitosis, **anaphase**, the centromeres split. The spindle fibres shorten, pulling the two halves of each centromere in opposite directions. One chromatid of each chromosome is pulled to each of the poles. Anaphase ends when the separated chromatids reach the poles and the spindle breaks down.

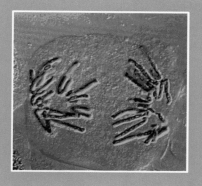

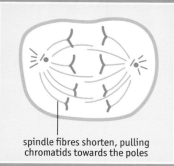

spindle fibres shorten, pulling chromatids towards the poles

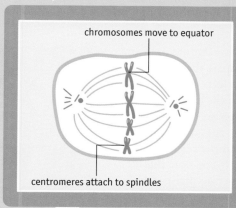

chromosomes move to equator

centromeres attach to spindles

METAPHASE

The next phase is called **metaphase**. The chromosomes' centromeres attach to spindle fibres at the equator. When this has been completed the cell has reached the end of metaphase.

A

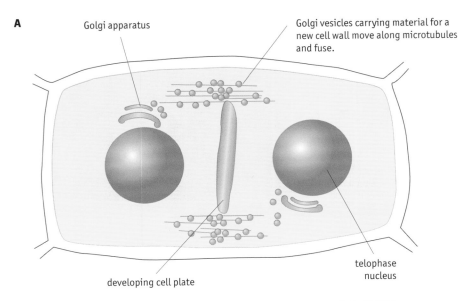

Golgi apparatus

Golgi vesicles carrying material for a new cell wall move along microtubules and fuse.

developing cell plate

telophase nucleus

B

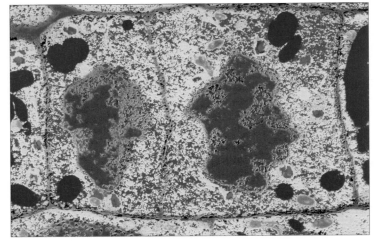

Figure 3.24 A The formation of the cell plate during cytoplasmic division of a plant cell. **B** Photomicrograph of a plant cell undergoing cytoplasmic division. Photo magnification ×6000.

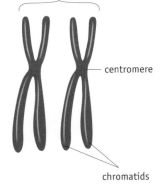

homologous chromosomes

centromere

chromatids

Figure 3.25 During prophase, each chromosome becomes visible as two strands, often called sister chromatids. The centromere joining the two chromatids may be in the middle of the strands or towards one end.

Q 3.16 Look at Figure 3.26 and compare the number of cells in telophase with the number in anaphase. What does this suggest to you about the relative lengths of time that these two phases take?

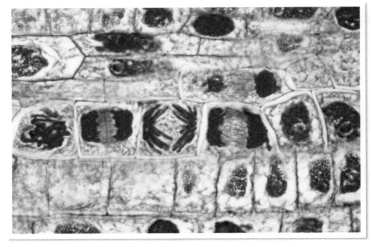

Figure 3.26 Cells undergoing mitosis. Magnification ×1170.

KEY BIOLOGICAL PRINCIPLE: WHY IS MITOSIS SO IMPORTANT?

Mitosis ensures genetic consistency, with daughter cells genetically identical to each other and to the parent cell. Each daughter cell contains exactly the same number and type of chromosomes as their parent cell. This is achieved by:

- DNA replication prior to nuclear division
- the arrangement of the chromosomes on the spindle and the separation of chromatids to the poles.

This genetic consistency is important in growth and repair, and also in asexual reproduction.

Growth and repair

Mitosis occurs in the growth of any organism as it develops from a single cell into a multicellular organism. It ensures that a multicellular organism has genetic consistency; all the cells in the body have the same genetic information. Some organisms can regenerate lost or damaged parts of their bodies using mitosis. Starfish can even re-grow a completely new body from a fragment (Figure 3.27A). Mitosis also allows old and damaged cells to be replaced with identical new copies.

Asexual reproduction

Many organisms reproduce without producing gametes. They grow copies of themselves by mitosis, producing offspring that are genetically identical to each other and to their parent. This form of reproduction is asexual, meaning 'without sex'.

Asexual reproduction occurs when the single-celled fungus, yeast buds. An outgrowth from the yeast cell, with DNA replication followed by mitosis, gives rise to a daughter cell once the bud detaches from the parent cell. *Hydra* also reproduce asexually by budding (Figure 3.27B). Repeated mitotic cell division at one point on the body of the hydra produces an outgrowth that develops and detaches to form a new

individual. When there is a shortage of food, both yeast and hydra can reproduce sexually, producing sex cells by meiosis. Asexual reproduction is common in plants too, where it is sometimes known as vegetative reproduction. The growth of new plants from tubers and bulbs are familiar examples (Figure 3.27C).

It is worth noting that there are organisms, such as mosses, liverworts and many flowering plants, that reproduce asexually at one stage in their life cycle and sexually at another.

Note: Prokaryotes, such as bacteria, do *not* carry out mitosis or meiosis, because they do not possess chromosomes. However, they do carry out binary fission, one cell splits into two identical cells.

Q 3.17 For each pair of statements, decide which one refers to asexual and which to sexual reproduction:

(a) one parent needed/two parents needed

(b) gametes/no gametes

(c) fertilisation/no fertilisation

(d) offspring show genetic variation/offspring genetically identical

(e) allows for rapid increase in population numbers/allows only slow increase in numbers

(f) unfavourable changes in the environment usually kill whole population/changes in the environment may only kill poorly adapted individuals

(g) effective for rapid colonisation of new areas/enables adaptation to changes in the environment.

Q 3.18 Suggest the advantage for yeast of entering meiosis when conditions are unfavourable rather than reproducing asexually.

A

B
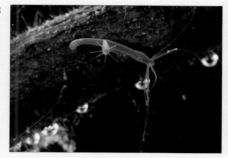
C

Figure 3.27 **A** Many organisms can regenerate parts of their bodies using mitosis. The crown-of-thorns starfish eats coral and has caused considerable damage to some coral reefs, including the Great Barrier Reef off Australia. Early attempts to control this starfish included chopping them up and throwing the bits back into the sea. Unfortunately, due to their ability to regenerate new individuals from many of the pieces, this just increased their numbers. **B** Freshwater *Hydra* attached to water weed. Notice the young hydra that is budding from the parent by asexual reproduction. Eventually, constriction at the point of attachment causes the release of the young hydra as a separate individual. **C** A hyacinth bulb is a highly modified underground shoot with fleshy leaves. Each of the shoot's buds can develop into a new plant.

Early embryonic development – stem cells

Cells in the early embryo

After a human zygote has undergone three complete cell cycles, it consists of eight identical cells. Each of these cells is said to be **totipotent** as it can develop into a complete human being. This is what happens when identical twins (or triplets or even quadruplets) form – such twinning can occur up to 14 days after conception. (These cells are called 'totipotent' as they have the 'potential' to develop into a 'total' individual.)

Q 3.19 In most animals, fertilisation is followed by a series of rapid cell divisions. In these first few divisions, the embryo divides without growing in size. Smaller cells are produced with each successive division. Large reserves of nutrients and extra cell contents that eggs accumulate allow the zygote to divide rapidly soon after fertilisation. Which phases of the cell cycle will be shortened to achieve this rapid division?

By five days after conception, a hollow ball of cells called the **blastocyst** has formed (Figure 3.28). The outer blastocyst cell layer goes on to form the **placenta**. The inner cell mass of 50 or so cells goes on to form the tissues of the developing embryo. These 50 cells are known as **pluripotent embryonic stem cells**. Each of these cells can potentially give rise to most cell types, though they cannot each give rise to all of the 216 different cell types that make up an adult human body. (The 'pluri' of 'pluripotent' is like the word 'plural'. It means that the cell can give rise to many cell types but, unlike the earlier totipotent cells, not all of them.

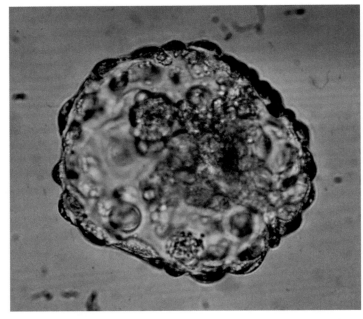

Figure 3.28 Five days or so after conception, a human zygote has divided to form a hollow ball of cells, the blastocyst. Between six and 12 days after fertilisation, the embryo implants in the wall of the uterus. Magnification ×250.

Cells become more differentiated

As the embryo develops into a multicellular body, the cells from which it is made become increasingly **differentiated**. Most of them lose the capacity to develop into a wide range of cells. Instead, they become increasingly specialised, functioning as a red blood cell, one of the cell types in bone, a plant xylem vessel or whatever (Figure 3.35, page 132). However, even in adults some cells retain a certain capacity to give rise to a variety of different cell types. These cells are known as **multipotent** stem cells. For example, neural stem cells can develop into the various types of cell found in the nervous system, while blood stem cells, located in bone marrow, can develop into red blood cells, platelets and the various sorts of white blood cells (macrophages, lymphocytes and so on). These are often called adult stem cells.

The presence of stem cells in an adult is not unique to humans, or even to mammals. For example, in the body walls of cnidarians, such as sea anemones and jellyfish, there are cells that on first sight appear to have no function. They do not help to defend the animal, feed it or protect it. They are just small, unspecialised cells.

Cnidarians have specialised cells that they use for paralysing their prey. These specialised cells work rather like harpoons and can only be fired once. They need to be constantly replaced and this is the role of the unspecialised cells, which divide and differentiate when more of the harpoon-like cells are needed.

Unlike animal cells, many plant cells remain totipotent throughout the life of the plant. Cell differentiation is irreversible in animals, but many differentiated plant cells can de-differentiate and then develop into a complete new plant. Gardeners have long known of this ability to regenerate identical whole plant clones from root, stem or leaf cuttings (Figure 3.29).

Totipotency of plant cells allows plants to be reproduced using plant tissue culture. Small pieces of a plant, known as explants, are surface-sterilised and then placed on a solid agar medium with nutrients and growth regulators. The cells divide to form a mass of undifferentiated cells known as a callus. By altering the growth regulators in the medium, cells of the callus can be made to differentiate to form small groups of cells that are very similar to plant embryos. These embryos develop into complete plants that are genetically identical clones. Plant tissue culture allows commercial growers to produce large numbers of genetically identical plants rapidly (Figure 3.30). Tissue culture is also important in plant biology research, plant breeding, genetic modification of plants and in the conservation of endangered plants.

Figure 3.29 Clones of the house plant *Streptocarpus* – common name: Cape primrose – are produced by planting leaf segments in compost.

Figure 3.30 Cereal plants being grown in test tubes from tissue cultures. Tissue culture techniques allow large-scale production of genetically identical, disease-free plants for horticultural and agricultural use.

WEBLINK

To find out how plant tissue culture and micropropagation are being used at Kew in the conservation of endangered plants, visit the Royal Botanic Gardens website.

ACTIVITY

In **Student Activity 3.11** you can demonstrate totipotency of plant cells using tissue culture techniques.

Use of human stem cells in medicine

Stem cells offer great hope to medicine, particularly in the area of regenerative medicine. This is the branch of medicine concerned with replacing, engineering or regenerating human cells, tissues or organs to achieve normal function. One day universal human donor cells may be produced that would provide new cells, tissues or organs for treatment and repair by transplantation. Embryonic stem cells may be the most suitable type of stem cells for this sort of treatment. Their potential to develop into any cell type offers the greatest flexibility for development; unlike adult stem cells, which are committed to developing only into certain cell types.

CHECKPOINT

3.5 Write a clear definition for each of the following terms used to describe stem cells: totipotent, pluripotent and multipotent.

Pluripotent stem cells for research and medicine can be isolated from so-called 'spare embryos'. These are produced in fertility clinics that carry out *in vitro* fertilisation, where the ovum is fertilised outside the body. Women undergoing this treatment are given drugs to make them superovulate, producing more eggs than are needed for infertility treatment. Many of the embryos end up being placed in women's wombs in the hope of enabling infertile couples to have children, but any additional embryos could be a source of embryonic stem cells. Currently, some of these embryos are donated for research purposes. The use of embryos in this way is controversial – see the section in Topic 2 about ethical concerns (Section 2.8).

Figure 3.31 Stem cells have been grown in culture. Researchers have successfully grown tissues, such as tendons, cartilage and heart value tissue from stem cells. Technically, though, this is difficult and the work is still at the research stage.

In the laboratory, embryos being used for research are allowed to grow to form blastocysts. At this point, the embryos are cultured for a further period of time to see if stem cells are formed. The stem cells are isolated from each embryo and the rest of the embryo (which could not develop further) is discarded. The stem cells are then cultured and used in research. In the future, they might be used to develop tissues that could be used for routine transplantation.

One problem with this approach is that even if scientists manage to get the stem cells to develop into the right sort of tissue, the tissue might end up being rejected by the immune system of the person given the transplant.

It might be possible to get round the problem of rejection either by using drugs that prevent the recipient from rejecting any transplanted organ or by using tissue-typing. Again, this approach has been around for many years and is widely used with conventional organ transplants.

An alternative and novel approach has been proposed and is generally referred to as **therapeutic cloning**, though the use of this term is regretted by many of the scientists working in the area. In therapeutic cloning, the patient needing a transplant would have one of their diploid cells removed – this could simply be a cell from the base of a hair or any other suitable tissue. This cell, or its nucleus, would then be fused with an ovum from which the haploid nucleus had been removed (Figure 3.32). The result would be a diploid cell rather like a zygote. This process is known as somatic cell nuclear transfer (a somatic cell is any diploid body cell).

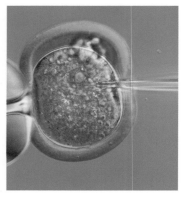

Figure 3.32 The ovum is held in place with a pipette. Its genetic material is removed and replaced with an adult cell nucleus.

This cell could then be stimulated to divide by mitosis in the same way as the cell that gave rise to Dolly the cloned sheep (see Figure 3.38, page 133). If all went to plan, after about five days a blastocyst would develop. Stem cells (Figure 3.33) could then be isolated from this and encouraged to develop into tissues. This procedure results in cell lines, and perhaps eventually organs for transplantation, which are genetically identical to the patient from whom the original diploid cell was taken.

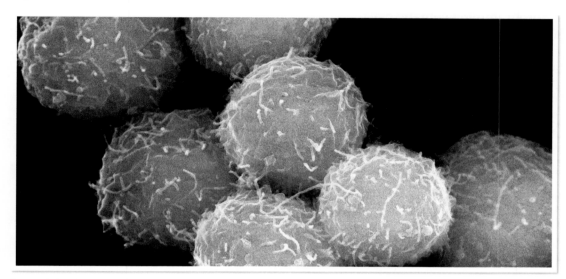

Figure 3.33 Somatic cell nuclear transfer would allow patient specific stem cells to be created which could be used in treatment.

In 2004, the UK Human Fertilisation and Embryology Authority (HFEA) gave permission for the use of therapeutic cloning in research. Somatic cell nuclear transfer is used to produce blastocysts from which human stem cells are extracted. These stem cells are used for research into diseases such as Parkinson's disease and motor neurone disease. The cloned embryos created have to be destroyed at 14 days old: it is against the law for cloned embryos to be used for reproductive purposes, even though this may technically be possible. The human ova come from consenting IVF patients, but such eggs are in limited supply.

In 2007, the HFEA decided to allow research using animal-human embryos. An animal ovum (e.g. from a cow) can be used, rather than a human ovum, so research is no longer limited by the supply of human ova. The nucleus of the animal egg is removed and a human nucleus is fused with the egg cell: the resulting cloned embryos are a source of stem cells for research.

Using adult stem cells

Many people have ethical objections to the use of embryonic stem cells and advocate the use of adult stem cells. Adult stem cells have been used for a long time in bone marrow transplants for treatment of leukaemia and related cancers. The adult stem cells move into the patient's bone marrow and produce healthy blood cells. Researchers report success in trials for stem cell treatments being developed for a range of degenerative conditions using patients' own stem cells. These include injection of stem cells into stroke patients' brains to replace damaged neurons and the injection of stem cells into the joints of patients with arthritis to repair damaged cartilage.

Adult stem cells have also been used to produce cells for transplantation (Figure 3.31). To treat patients with burns, the patient's skin stem cells from an unburned area have been cultured. The cells multiply, forming a sheet of cells that cover the bottom of the culture flask. These cells can be transplanted to the burned area of the patient giving a skin surface, although because the adult stem cells cannot differentiate, the layer of skin cells do not contain structures such as sweat glands. June 2008 saw the first experimental transplant involving a patient's own stem cells. A patient received a trachea created using a donor windpipe stripped of the donor's cells to leave the cartilage skeleton, with the patient's stem cells then grown on its surface. In 2011, the first synthetic windpipe built up with the patient's own stem cells grown on a plastic scaffold was transplanted into a patient (Figure 3.34). Stem cells have also been used in the production of blood vessels and even a nose for transplantation. However, the isolation and culturing of stem cells is difficult and costly, so the culturing of stem cells lines for use would be a more practical option. If whole organs are one day grown from stem cells – and this is a very long way off – it has been estimated that perhaps only 20 stem cell lines would be needed to provide transplants for 90% of the UK population.

Reprogramming somatic cells

In 2006, Japanese researchers, Kazutoshi Takahashi and Shinya Yamanaka, reported the successful reprogramming of mouse somatic (body) cells to make them pluripotent. These cells are known as induced pluripotent stem cells (iPSCs). Takahashi and Yamanaka, and other research groups, have since produced human iPSCs that resemble human embryonic stem cells. Intense research effort is focused on the development of this cell reprogramming. It would overcome the problems of cell rejection and address ethical concerns with the use of embryonic stem cells. Working in animal models, iPSC treatment has been shown to be effective for a range of diseases, including sickle cell anaemia, Parkinson's disease, haemophilia and ischaemic heart disease.

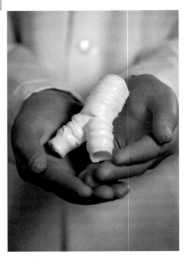

Figure 3.34 In a pioneering operation, a patient whose trachea was blocked by a tumour the size of a golf ball had the affected section replaced with one created from his own stem cells and a type of plastic.

> **WEBLINK**
>
> You can read more about Yamanaka's work on the Nobel Prize website. He shared the 2012 Nobel Prize in Physiology or Medicine with John B. Gurdon for the discovery that mature cells can be reprogrammed to become pluripotent. In a classic experiment in 1962, Gurdon discovered that the specialisation of cells is reversible: a frog egg cell that had its nucleus replaced with the nucleus from a mature intestinal cell developed into a normal tadpole.

As it is possible to maintain embryonic and induced pluriopotent stem cells in tissue culture (Figure 3.31), they are being used for research into both human development and disease. For example, they can be used to explore how genes trigger the onset of development, such as organ formation. Stem cell research may help us to understand how cancer cells develop and how certain birth defects occur. Stem cells could also be used to provide a source of normal human cells of virtually any tissue type for use in screening new drugs.

Ethical concerns about the use of stem cells

Just about everyone agrees that there are no ethical objections to using multipotent stem cells derived from adults: for example, from adult bone marrow. The problem is that scientists believe that these stem cells are likely to be less valuable for research and for the development of new treatments than pluripotent stem cells, which can only be derived from human embryos. The development of iPSCs has led some people to suggest that this solves the ethical concerns of using embryonic stem cells. However, there is debate about how similar embryonic and induced pluripotent stem cells really are. It remains unclear whether iPSCs could eventually replace embryonic stem cells in research and future treatments. Ongoing research on all types of stem cells will be necessary to decide this. Even if adult stem cells or iPSCs are used for new research, they may be of little use unless a better understanding is gained of how they specialise. This understanding may only come from embryonic stem cells.

Different people see the status of the human embryo very differently. As the official UK government committee set up to report on human stem cell research – with the Chief Medical Officer, Professor Liam Donaldson, in the chair – said in the year 2000:

> *A significant body of opinion holds that, as a moral principle, the use of any embryo for research purposes is unethical and unacceptable on the grounds that an embryo should be accorded full human status from the moment of its creation. At the other end of the spectrum, some argue that the embryo requires and deserves no particular moral attention whatsoever. Others accept the special status of an embryo as a potential human being, yet argue that the respect due to the embryo increases as it develops and that this respect, in the early stages in particular, may properly be weighed against the potential benefits arising from the proposed research.*

> **WEBLINK**
>
> Visit the Euro Stem Cell website for lots of information and fact sheets about stem cells.

The role of the regulatory authorities

Much scientific work can simply be applied usefully without the need for any regulatory authorities to get involved. However, there are times when new advances in society raise various ethical or other concerns and in these cases, regulation may be needed.

In the UK, the ultimate regulatory authority is parliament. New laws are passed by majorities in both the House of Commons and the House of Lords. Then the wording of these laws is interpreted by the courts.

In matters of science, those responsible for making new laws can receive advice from a number of sources. For example, there are 'Select Committees' in both the Lords and the Commons. These are made up of peers of the realm (in the Lords) and MPs (in the Commons), but they are advised by specialist advisors who have particular expertise in the subject in question.

There are also various Advisory Committees. For example, the Advisory Committee on Novel Foods and Processes advises the government as to whether new foods (including foods made from genetically modified ingredients) should be permitted. Its members are mainly independent scientists, but also include an ethicist.

Scientists make an objective interpretation of the scientific evidence. This scientific knowledge, and public consultation, help the regulatory authorities make decisions about the applications and implications of the science in the context of society.

In the UK, the Human Fertilisation and Embryology Authority (HFEA) regulates research on human embryos. Until 2001, UK law only allowed the use of human embryos where the HFEA considered their use to be necessary or desirable:

- to promote advances in the treatment of infertility
- to increase knowledge about the causes of congenital disease
- to increase knowledge about the causes of miscarriage
- to develop more effective methods of contraception
- to develop methods for detecting gene or chromosome abnormalities in embryos before implantation.

On 22 January 2001, peers in the UK House of Lords voted by 212 to 92 votes to extend the purposes for which research on human embryos is allowed. (A vote in the House of Commons had already gone the same way in December 2000 by 366 to 174 votes.) So-called 'spare' embryos from *in vitro* fertilisation treatment can now be used as a source of embryonic stem cells for the purpose of research into serious disease. On 27 February 2002, this was extended to include fundamental research necessary to understand differentiation and de-differentiation of cells.

These 2001 and 2002 extensions have led the HFEA to permit new areas of work, including its 2007 decision to allow research using animal-human embryos. However, this whole area of work is very controversial and it is possible that those opposed to such work may attempt to take the HFEA to court. Ultimately, it is parliament that lays down the laws and it is the courts that decide whether or not laws have been broken.

 ACTIVITY

In **Student Activity 3.12** you can read some quotes from those who are for and against the use of stem cells, and decide what your own position is on the issue.

3.3 How development is controlled

In Figure 3.35 you can see an early stage in the development of a mouse. As cells divide after fertilisation they become specialised for a small number of functions (Figure 3.36). As you probably already know, specialised cells often work together as tissues in organs.

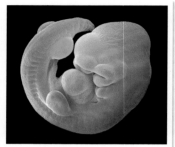

Figure 3.35 A mouse during an early stage in its development.

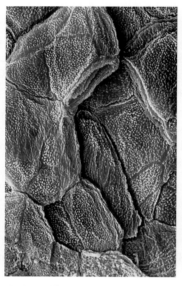
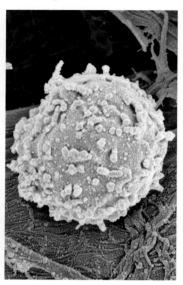

Figure 3.36 Although these specialised cells look different to one another, they have all been produced by mitosis from a common ancestor so we know they all share some common features and must all contain a complete complement of chromosomes with the same genetic information.

We now take it for granted that the nucleus has a role in controlling the development of the individual cell and the whole multicellular organism's phenotype. This was first shown in classic experiments using giant algal cells, completed in 1934 by a Danish biologist named Joachim Hammerling (Figure 3.37).

1

hat

stalk

rhizoid containing nucleus

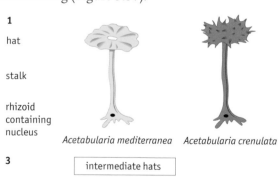

Acetabularia mediterranea *Acetabularia crenulata*

2

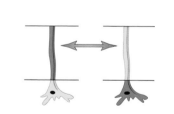

Hats are removed and the stalks are swapped.

3

intermediate hats

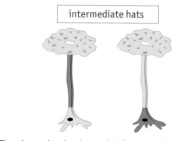

The plants develop hats with features of both species. What controls development of the hats – the nucleus or chemicals in the cytoplasm?

4

If the intermediate hats are removed, new ones grow that correspond to the nucleus in the rhizoid. This demonstrates the role of the nucleus and chemical messengers in the development of the cell

Figure 3.37 Is it the nucleus or the cytoplasm that controls development in *Acetabularia mediterranea* and *A. crenulata*? These results showed that the nucleus and chemical messenger(s) moving through the cytolasm control development of the cell.

Q 3.20 Suggest what the chemical messengers are that control the growth of hats that correspond to the nucleus in the rhizoid, as shown in part 4 of Figure 3.37.

In Topic 2, we saw the role of genes in determining the structure and function of cells through the genetic code in the DNA transcribed and translated in protein synthesis. However, a problem arises when one considers that all the cells in a multicellular organism are derived from a single cell and through DNA replication and mitosis, they all contain the same genetic information, except for any mutations. It raises the question: how is it possible that there are lots of different types of specialised cell yet they all contain the same genetic information?

Q 3.21 In 1997, Ian Wilmut and his colleagues at the Roslin Institute in Scotland successfully cloned an adult sheep. They transplanted the nucleus from a mammary gland cell of one adult sheep into another sheep's unfertilised ovum from which the nucleus had been removed. The diploid cell that formed developed into an embryo which was implanted into another adult sheep (Figure 3.38). The surrogate mother gave birth to a lamb who soon became the world famous 'Dolly'.

How did the creation of Dolly support the idea that all the genetic information for making a complete organism is present in every single cell, including those in an adult with specialised functions?

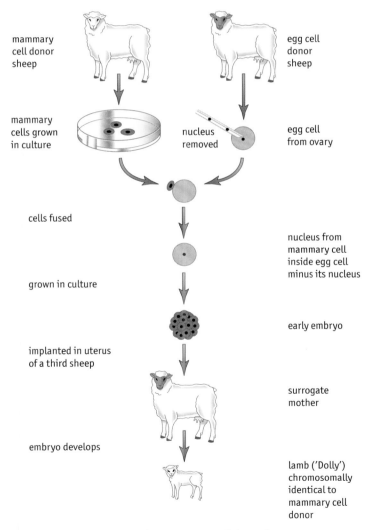

Figure 3.38 Dolly's genes were identical to those of the mammary cell donor sheep.

Problems with cloning

Animal cloning is not yet as easy as you might think from the description in Q3.21. In fact, most developmental biologists (including Ian Wilmut) feel that the process is not safe to try on humans. Many attempts to produce live-born animals by cloning have been unsuccessful and a high proportion of the animals produced by cloning have health problems. Dolly herself was the sole success from 277 attempts at cloning and this low success rate is still typical. Even Dolly may not have been completely 'normal' as she developed arthritis at a fairly young age for a sheep and had to be put down.

The reasons why cloning mammals from adult nuclei has such a low success rate are not yet fully understood and there are probably several different contributing factors. One of these is that the DNA in an adult cell nucleus has been programmed into a particular type of cell (e.g. a skin cell). When transferred to an ovum, the nucleus may not be able to reprogram its DNA quickly enough to be able to switch on all the different genes required for normal development.

Different genes are expressed

As the embryo develops, cells differentiate: they become specialised for one function or a group of functions. The structure and function of each cell type is dependent on the proteins that it synthesises. Salivary gland cells make salivary amylase, an enzyme for digesting starch, while red blood cells contain the protein haemoglobin, an oxygen-carrying pigment. Amylase and haemoglobin are both proteins coded for by the cells' genes. How is it that these cells, and only these cells, produce these proteins even though every cell contains the complete genome, including the codes for both amylase and haemoglobin? Different specialised cells must only be expressing some of their genes.

Work by Igor Dawid and Thomas Sargent demonstrated that different genes are expressed in different cells. They extracted messenger RNA (mRNA) from undifferentiated and differentiated frog cells (Figure 3.39). Complementary DNA (cDNA) strands were produced for all the mRNA in the differentiated cells using an enzyme called reverse transcriptase, which, as its name suggests, reverses transcription, making DNA from mRNA. These cDNA strands were mixed with the mRNA from the undifferentiated cell. Complementary strands of cDNA and mRNA combined to produce double strand hybrids. When these hybrids were separated out, there remained a range of cDNA strands that had not been hybridised: the two cells were expressing some of the same genes, but also some different genes.

Switching genes on

Cells become specialised because only some genes are switched on and produce active mRNA that is translated into proteins within the cell.

The role of the epigenome

The **epigenome** influences which genes can be transcribed in a particular cell. The epigenome should not be confused with the genome, which is all the DNA containing a full set of genes. DNA is wrapped around histone proteins (Figure 3.40) and both the DNA and histones have chemical markers attached to their surface. These chemical markers make up the epigenome.

The attachment of, for example, methyl groups ($-CH_3$) to the DNA of a gene, usually to cytosine, prevents transcription to mRNA, by stopping the RNA polymerase binding. The modification of histones by addition of, for example, methyl or acetyl groups, affects how tightly the DNA is wrapped around the histone. When wound tightly, the genes are inactive: they cannot be transcribed to mRNA (Figure 3.40). The gene therefore cannot make protein; it is 'switched off'.

In cells of a particular type of tissue, such as the pancreas, all the genes for the pancreas are active (for instance, the gene for making the enzyme trypsin), but all the genes specific to other tissues, such as the liver and kidneys, are switched off. During DNA replication, the epigenetic markers are copied with the DNA so that when a pancreas cell divides it forms another pancreas cell: the correct set of genes remain active. Although all the active genes may not all be transcribed continuously.

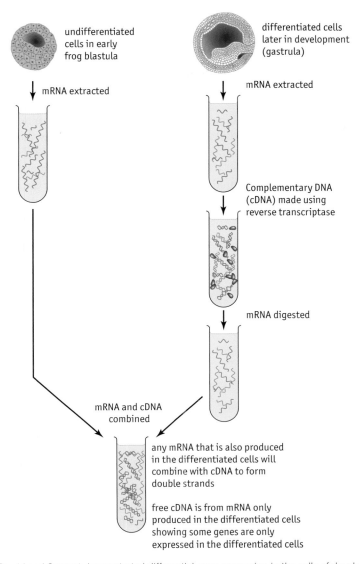

undifferentiated cells in early frog blastula

↓ mRNA extracted

differentiated cells later in development (gastrula)

↓ mRNA extracted

Complementary DNA (cDNA) made using reverse transcriptase

mRNA digested

mRNA and cDNA combined

any mRNA that is also produced in the differentiated cells will combine with cDNA to form double strands

free cDNA is from mRNA only produced in the differentiated cells showing some genes are only expressed in the differentiated cells

Figure 3.39 How Dawid and Sargent demonstrated differential gene expression in the cells of developing frogs.

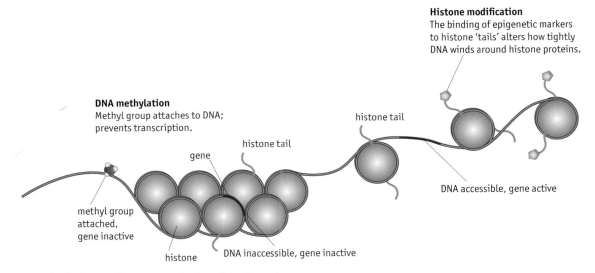

Histone modification
The binding of epigenetic markers to histone 'tails' alters how tightly DNA winds around histone proteins.

DNA methylation
Methyl group attaches to DNA; prevents transcription.

histone tail

histone tail

DNA accessible, gene active

gene

methyl group attached, gene inactive

histone

DNA inaccessible, gene inactive

Figure 3.40 Control of gene expression by epigenetic markers.

Switching on an individual gene – the lac operon model

The French geneticists Jacob and Monod were the first to propose a theory for the control of gene expression in the early 1960s. They studied the control of genes in the prokaryote *Escherichia coli*. These bacteria only produce the enzyme **β-galactosidase** to break down the carbohydrate lactose when it is present in the surrounding medium. This enzyme converts the disaccharide lactose to the monosaccharides glucose and galactose.

When lactose is not present in the environment, a lactose repressor molecule binds to the DNA and prevents the transcription of the β-galactosidase gene. Figure 3.41 shows how the lactose repressor binds to the operator and stops the β-galactosidase gene being expressed. The RNA polymerase cannot bind to the DNA promotor region. If lactose is present in the environment it binds to the repressor, the repressor molecule is prevented from binding to the DNA and the β-galactosidase gene is expressed. mRNA coding for β-galactosidase is transcribed, and translation of this mRNA produces the enzyme.

The operator and genes associated with it are known as an operon, in this case the lac operon.

ACTIVITY

In **Student Activity 3.14** you can see how a gene is switched on by the presence of a substrate.

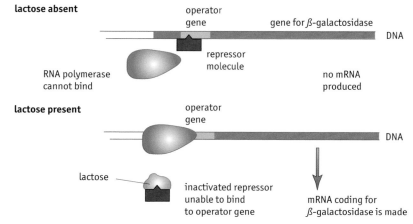

Figure 3.41 Control of the β-galactosidase gene in the bacterium *Escherichia coli*. The RNA polymerase must bind for transcription to start.

What switches transcription of an individual gene on or off in eukaryotes?

Genes in uncoiled, accessible regions of the eukaryote DNA can be transcribed into mRNA. The enzyme RNA polymerase binds to a section of the DNA adjacent to the gene to be transcribed. This section is known as the promoter region. Only when the enzyme has attached to the DNA will transcription proceed. This process is outlined in Figure 3.42. The gene remains switched off until the enzyme attaches to the promoter region successfully. The attachment of a regulator protein is usually also required to start transcription.

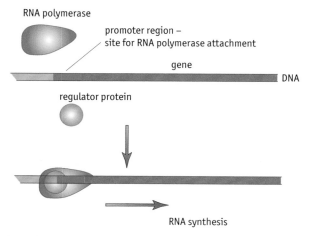

Figure 3.42 DNA transcription will start only when the RNA polymerase has attached to the promoter region of the gene that is to be transcribed.

Transcription of a gene can be prevented by protein repressor molecules attaching to the DNA of the promoter region, blocking the attachment site. In addition, protein repressor molecules can attach to the regulator proteins themselves, preventing them from attaching. In either case, the gene is switched off: it is not transcribed within this cell.

When gene expression goes wrong

The importance of differential gene expression in producing specialised cells is highlighted when things go wrong. Here is an example.

Growing bones in the wrong places

Diana is a baby whose big toes are too short and are curved inwards towards her other toes. When her parents took her to a specialist they were told that this can be the first sign of a rare inherited disease called FOP (*fibrodysplasia ossificans progressiva*). FOP is characterised by the growth of bones in odd places, such as within muscle and connective tissue (Figure 3.43). By early adulthood, the abnormal bone growth often leads to the 'freezing' of all major joints of the backbone and limbs so they cannot move. Any local trauma, even injection of medicines into muscle, can cause bone formation in people with FOP.

What causes FOP?

FOP is an inherited condition caused by a gene mutation. Bone cells are normally only produced in growing limbs and in other places where the skeleton develops. Here, genes are expressed that produce all the proteins needed to become a specialised bone cell. In people with FOP, one of these genes is not switched off in white blood cells. When tissue is damaged, for example when having an injection, the body's white blood cells move to the site of the damage. The white cells produce the protein, which diffuses into the surrounding muscle cells. The protein causes the muscle cells to start expressing other genes that turn them into bone cells.

How are cells organised into tissues?

Specialised cells can group themselves into clusters, working together as a tissue. Cells have specific recognition proteins, also known as adhesion molecules, on their cell surface membranes. Adhesion molecules help cells to recognise other cells like themselves and stick to them. A small part of each recognition protein is embedded in the cell surface membrane; a larger part extends from the membrane. This exposed section binds to complementary proteins on the adjacent cell (Figure 3.44). The particular recognition proteins synthesised by a cell determine which cells it can and cannot attach to. If cells from different tissues are separated and then mixed together, they reform into the tissues as the recognition proteins bind.

EXTENSION

In **Student Extension 3.5** read about how one X chromosome is inactivated in every female mammalian cell.

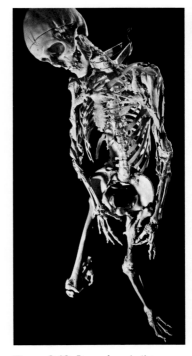

Figure 3.43 Bones form in the wrong places in people with the rare inherited disease called FOP. By their early 20s, most patients are confined to wheelchairs. Starvation used to result from the freezing of the jaw, and pneumonia can occur due to the lungs or diaphragm becoming fixed to the chest wall. Surgery can make the condition worse by causing more bone growth. There are no proven treatments that prevent progression of the disease.

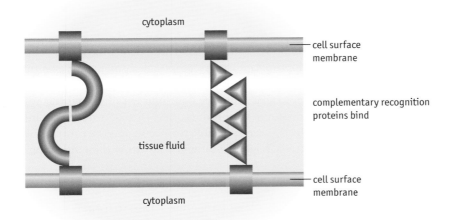

Figure 3.44 Recognition proteins synthesised by cells determine which cells work together in tissues and organs.

In the human embryo, cells begin to form tissues as they start their specialised functions. At this point, the genes coding for the recognition proteins are switched on. Reorganisation of tissues during development occurs through changes in the type of recognition proteins produced by the cells.

In tissues, cells also interact with the extracellular matrix, a network of molecules secreted by cells. In some tissues, the extracellular matrix is a major component of the tissue: for example, in cartilage.

Table 3.1 summarises the levels of organisation within organisms.

cell	In multicellular organisms, cells are specialised for a particular function. For example, muscle cells and epithelial cells.
tissue	A group of specialised cells working together to carry out one function. For example, muscle cells combining to form muscle tissue and epithelial cells forming epithelial tissue.
organ	A group of tissues working together to carry out one function. For example, muscle, nerve and epithelium work together in the heart.
organ systems	A group of organs working together to carry out a particular function. For example, the circulatory system.

Table 3.1 Summary of the levels of organisation within organisms.

Q 3.22 State **a** the level of organisation and **b** the function performed by each of the following examples:

- xylem vessel in a flowering plant
- leaf
- kidney
- neurone.

Gene expression and development

The precise sequence of transcription and translation of genes determines the sequence of changes during development. The epigenome helps to control the change from a single-celled zygote to a fully formed adult, containing a variety of tissues and organs. During development, epigenetic changes will bring about specialisation of the cell. Signals from inside and outside the cell result in changes to the epigenome that alters the genes transcribed at specific times and locations. Copying of the epigenome during DNA replication ensures that the changes occurring during development are passed on to new cells. By studying the embryonic development of model animals and plants, researchers have shown how differentiated gene expression determines the structures produced during development.

Master genes

In fruit flies, *Drosophila*, once the main body segments have been determined, the cells in each segment become specialised for the appropriate structures in that segment: for example, legs, antennae and wings. Master genes control the development of each segment. These genes were discovered by looking at mutations that cause the development of the wrong appendage for that segment (Figure 3.45). The master genes produce mRNA that is translated into signal proteins. These proteins switch on the genes responsible for producing the proteins needed for specialisation of cells in each segment. In Topic 7, you will study how these signal proteins switch on genes in detail.

The ABC of flowering plants

When a plant starts to flower, cells in a meristem become specialised to form the organs that make up the flower. Flowering of *Arabidopsis thaliana* has been used to model what is happening when plants change from vegetative growth to reproductive development. Most hermaphrodite flowers – those that have male and female structures – contain four sets of floral organs: sepals, petals, male stamens and female carpels. These are arranged in concentric whorls (Figures 3.46 and 3.47).

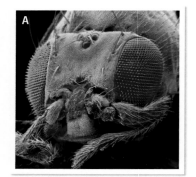
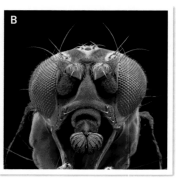

Figure 3.45 A In this mutant fruit fly, legs have developed where the antennae should be. **B** The head of a normal fruit fly. Magnification ×50.

The expression of genes in cells across the meristem determines which structures will form. Three genes determine which type of specialised organ will be produced in each area of the meristem (Figure 3.45). Where only gene A is expressed, sepals form and where only gene C is expressed, carpels form. Petals form where A and B genes are expressed, and stamens form where B and C genes are expressed. As with the master genes in fruit flies, these genes produce mRNA that codes for signal proteins that switch on appropriate genes. Synthesis of proteins coded for by these genes results in development of specialised cells in each floral organ.

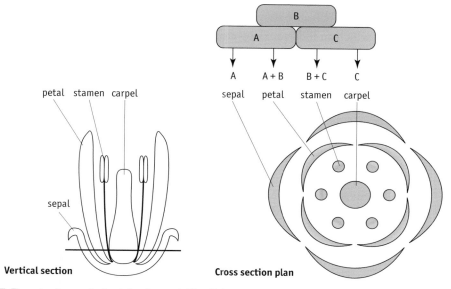

Figure 3.46 *Arabidopsis thaliana* (wall cress), a member of the mustard family, is used extensively in genetic and molecular research.

Figure 3.47 The role of genes in floral development. The ABC genes active in the formation of the flower are shown at the top, with their corresponding structures highlighted on the cross-section plan of the flower below.

Q 3.23 The role of the three master genes that control flower development was determined by looking at plants with mutations. One mutation causes a plant to only produce genes A and B.

(a) Name the floral structures that would form within the flowers of this plant.

(b) Describe how these would be arranged in the flower (a diagram would be useful).

⚙ ACTIVITY
In **Student Activity 3.15** you construct flowers using the ABC model.

✓ CHECKPOINT
3.6 Write a short paragraph that summarises how cells become specialised for their different structures and functions.

DID YOU KNOW?

Programmed cell death (apoptosis)

It is an amazing fact that all mammal cells have a 'self-destruct' programme. During development, an animal must lose some cells by programmed cell death, called apoptosis. There is a small group of cell 'suicide' genes and when they are expressed this causes the nucleus and cytoplasm to fragment.

Genes that prevent death are expressed in most cells and during development, so cells have to switch on their 'suicide' genes in order to die. This explains why cell death does not happen throughout the embryo. Programmed cell death occurs in particular places, such as between where the fingers will be. In the embryo brain, many millions of extra cells are produced, at least half of which die by controlled cell death.

3.4 Genes and environment

Is it all in the genes? Nature and nurture

The characteristics of an organism, such as its size, shape, blood group or sex, are known as its **phenotype**. Differences in phenotype between the members of a population are caused by differences in:

● genetic make-up or **genotype**

● the environment in which an individual develops.

Some characteristics are controlled almost completely by the organism's genotype, with the environment having little or no effect. For example, a person's blood group (group A, B, AB or O) is controlled by the genes they inherit. The genes code for glycoprotein on the surface of their red blood cells. A person's blood group is not affected by the environment in which they develop. Such characteristics are controlled by genes at a single locus and show **discontinuous variation**. They have phenotypes that fall into discrete groups with no overlap (Figure 3.48A). In garden pea plants, unlike humans, height also shows discontinuous variation.

Characteristics that are affected by both genotype and environment often show **continuous variation**. Human height is a characteristic showing continuous variation. This means that a person can be any height within the human range. The most common height will be somewhere mid-way between the extremes of the range. If a graph is drawn showing the frequency distribution of the different height categories it will be bell-shaped (Figure 3.48B). Characteristics that show continuous variation are controlled by genes at many loci, known as polygenic inheritance. They are also controlled by the environment, either directly or by influencing gene expression.

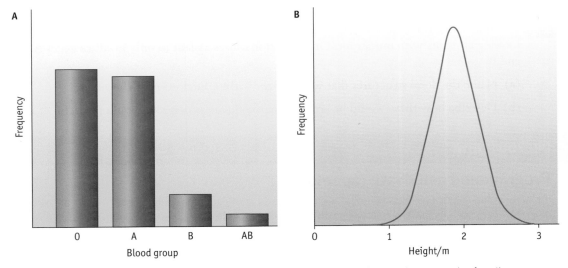

Figure 3.48 A Human blood groups show discontinuous variation. **B** Height in humans is an example of continuous variation.

KEY BIOLOGICAL PRINCIPLE: POLYGENIC INHERITANCE

In monohybrid inheritance, each locus is responsible for a different heritable feature. For example, one gene might be coding for the colour of a flower with another gene coding for the shape of the petals. However, inheritance of most characteristics does not follow simple Mendelian rules of inheritance, but involves interaction of alleles at many loci. When a number of genes are involved in the inheritance of a characteristic, rather than just one, we call the pattern of inheritance **polygenic**.

In many human characteristics that show polygenic inheritance, such as height and skin colour, the alleles clearly have additive effects. With conditions such as diabetes, coronary heart disease, Alzheimer's disease, schizophrenia and some cancers, several genes may confer a *susceptibility* to the condition, with environmental factors also contributing. Two people who inherit the same susceptibility may not both develop the illness: it will depend on environmental factors such as diet, exposure to toxins and stress. Such factors act as triggers to bring about the symptoms of disease. Conditions where several genetic factors and one or more environmental factors are involved are said to be **multifactorial.**

The degree of similarity between identical twins is a measure of the influence of the genes on that characteristic. 99% of identical twins have the same eye colour and 95% the same fingerprint ridge count. Where the environment has a greater effect, the similarity falls. One study found that if you are an identical twin and you have Alzheimer's disease, your twin sibling (brother or sister) has a 40% chance of having Alzheimer's disease. However, if you have Alzheimer's disease as a non-identical twin, your sibling has only a 10% chance of having Alzheimer's disease. This suggests that there is a significant but not inescapable genetic basis to Alzheimer's disease. Both genes and the environment influence the development of the disease.

How polygenic inheritance works

In any introductory course on genetics, it is common for eye colour to be used as an example of monohybrid inheritance – a single locus with brown dominant to blue. This is not entirely the case. Eye colour is an example of polygenic inheritance; alleles at several loci control eye colour. Eye colour ranges from blue to brown, depending on the amount of pigment in the iris. The pigment absorbs light so brown eyes appear dark. Blue eyes have little pigment, so light reflects off the iris.

Let us say three loci are involved in the inheritance of this characteristic, each with alleles **B** and **b**. **B** adds pigment to the iris and **b** does not. If all three loci were homozygous for the allele **B**, the person's genotype would be **BB BB BB**. The additive effect would produce a dark brown iris, whereas **bb bb bb** would add no pigment to the iris, making it pale blue. A range of possible genotypes and phenotypes are possible between these two extremes, according to how many alleles add brown pigment, as shown below.

Number of alleles adding brown pigment	Example of genotype	Eye colour
6	BB BB BB	very dark brown
5	BB BB Bb	dark brown
4	BB BB bb	medium brown
3	BB Bb bb	light brown
2	BB bb bb	deep blue
1	Bb bb bb	medium blue
0	bb bb bb	pale blue

Table 3.2 Eye colour is a polygenic characteristic.

The greater the number of loci involved, the greater the number of possible shades.

If a pale blue-eyed woman has children with a very dark brown-eyed man they will have light brown-eyed children as shown below.

	Mother	Father
Parental phenotypes	Pale blue	Very dark brown
Parental genotypes	**bb bb bb**	**BB BB BB**
Gametes	**bbb**	**BBB**
Offspring genotypes	**Bb Bb Bb**	
Offspring phenotype	Light brown	

Q 3.24 A deep blue-eyed woman (**Bb bb bB**) has a child with light brown eyes. Her medium blue-eyed partner (**Bb bb bb**) suspects that he is not the father. Copy and complete the Punnett square below and describe how you would use it to explain to him that he could be the father.

	Mother	Father
Parental phenotypes	Deep blue	Medium blue
Parental genotypes	**Bb bb bB**	**Bb bb bb**
Possible gametes	**Bbb, BbB, bbb, bbB**	**Bbb, bbb**

		Gametes from the father	
		Bbb	**bbb**
Gametes from the mother			

Height, weight and skin pigmentation all involve polygenic inheritance. Each allele has a small effect on the characteristic and the effects of several alleles combine to produce the phenotype of an individual. Suppose that only two genes were involved in the determination of height. The homozygous recessive genotype, **aabb**, might give a height of 20 cm above a baseline of 140 cm for adult women and 150 cm for adult men. In other words, the recessive alleles (**a** and **b**) each contribute 5 cm to the height. The dominant alleles, **A** and **B**, each contribute 10 cm to the height – so the homozygous dominant **AABB** would give a height of 40 cm above our baselines. **AaBb** would add 30 cm to the height.

If two heterozygotes were crossed, there would be a range of phenotypes as shown below.

Height above baseline/cm	20	25	30	35	40
Number of offspring with that height	1	4	6	4	1

Table 3.3 Range of phenotypes from crossing two AaBb heterozygotes.

Q 3.25 **(a)** Sketch what a graph of the data from Table 3.3 would look like.

(b) What can you conclude about height from this graph?

(c) For the cross between two heterozygotes, **i** state the parental phenotypes and possible gamete genotypes, and **ii** draw up a Punnett square to show the possible offspring genotypes and phenotypes.

If instead of two loci there were several, the number of height phenotypes would increase. The greater the number of loci, the greater the number of height classes and the smaller the differences between classes. If plotted, this would produce a smooth bell-shaped curve, typical of continuous variation (see Figure 3.48B).

Traits such as height and skin colour are not just the result of an individual's genotype. Their environment also has an influence. For example, diet has an effect on a person's height. A poor diet may prevent a person reaching their full height as predicted by their genotype. Their genotype gives their *potential* height. Their genotype in combination with their environment determines their *actual* height. The result is that there are no clear height classes, but a continuous variation in height between the extremes of the range.

ACTIVITY

In **Student Activity 3.16** discover how many genes can affect a single characteristic.

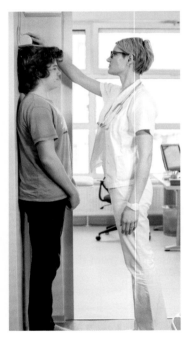

Figure 3.49 People are getting taller.

Gene and environment interactions

There are countless examples of genes and environment interacting to produce an organism's phenotype. Some are very familiar, such as skin and hair colour; others are less so. We will take a detailed look at some examples and consider how the environment might affect gene expression.

Height

Have you noticed that each human generation seems to be a bit taller than the last (Figure 3.49)? The average human height in industrialised countries has risen each generation over the past 150 years. People in the UK now average about 8 cm taller than they did in 1850. How can this be explained?

Before reading on, list as many reasons as you can for the increase in human height over recent generations.

The following are some possible answers. You may have thought of others.

1 There is some evidence that taller men have more children, which would result in a gradual change in the genetic make-up of the population.

2 Greater movements of people have resulted in less inbreeding, leading to taller people.

3 Better nutrition, especially increased protein, has resulted in greater growth of children.

4 Improved health, with a reduction in infectious diseases, has occurred through improved sanitation, clean water supplies, vaccination and antibiotics.

5 The end of child labour has allowed more energy to go into growth.

6 Better heating of houses and better quality clothing reduces the amount of energy needed to heat the body, so again more energy can go into growth.

ACTIVITY

Student Activity 3.17 lets you see if the difference in heights over the years can be observed.

Q 3.26 Which of these reasons for change in height are due to the environment and which are the result of genotype?

It is widely accepted that a person's height is determined by an interaction between the effects of their genes for height and environmental influences, such as diet. A person may have genes for being tall, but not achieve his or her full potential height because of malnutrition. We do not know for certain which of the possible reasons for change in height is the most important. However, most people think that better diet is the most significant factor.

Hair colour

Differences in hair colour are largely genetically determined, due to variation in the amount and type of pigment the hair contains. But the environment can influence hair colour in some cases.

Making melanin

The dark pigment in skin and hair is called **melanin**. Melanin is made in special cells, melanocytes, found in the skin and at the root of the hair in the follicle. These are activated by **melanocyte-stimulating hormone** (**MSH**). There are receptors for MSH on the surface of the melanocyte cells. The melanocytes place melanin into organelles called melanosomes. The melanosomes are transferred to nearby skin and hair cells where they collect around the nucleus, protecting the DNA from harmful UV light. People with more receptors have darker skin and hair; so, they have more protection against sunburn. Ultraviolet (UV) light increases the amount of MSH and also of MSH receptors, making the melanocytes more active (Figure 3.50) and causing the skin to darken. Hair does not appear darker because although more melanin is produced, UV light causes chemical and physical changes to melanin and other proteins in hair cells. Hair lightens due to destruction of the melanin by UV light.

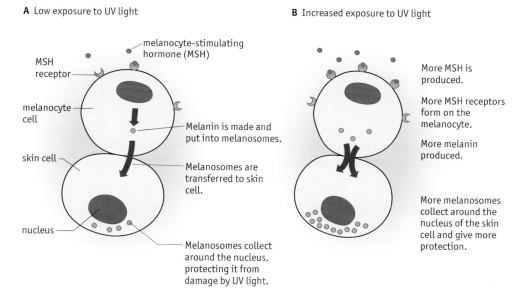

A Low exposure to UV light

- melanocyte-stimulating hormone (MSH)
- MSH receptor
- melanocyte cell
- Melanin is made and put into melanosomes.
- skin cell
- Melanosomes are transferred to skin cell.
- nucleus
- Melanosomes collect around the nucleus, protecting it from damage by UV light.

B Increased exposure to UV light

More MSH is produced.

More MSH receptors form on the melanocyte.

More melanin produced.

More melanosomes collect around the nucleus of the skin cell and give more protection.

Figure 3.50 The formation of melanin in melanocytes.

Seasonal colour change

Arctic foxes have brown fur in summer and white fur in winter (Figure 3.51). They must have the genes for making brown fur (which contains melanin) all the time, so how can white fur be made?

The white winter coat is actually grown during the summer. It grows under the brown summer coat and is revealed when the summer coat is moulted in autumn. The foxes produce fewer MSH receptors in the summer. Without these receptors, MSH has no effect and no melanin is made in the hair follicles.

Q 3.27 Explain why MSH has no effect on the developing hairs in summer, meaning that the coat grows white.

Q 3.28 Explain why it is surprising that white fur is able to grow in the summer.

EXTENSION

In **Student Extension 3.6** find out how both genotype and environment influence sex determination.

Figure 3.51 How does the arctic fox change colour?

White with dark tips

To make melanin, animals use an enzyme called **tyrosinase**. Tyrosinase catalyses the first step along a chemical pathway, changing the amino acid tyrosine into melanin. Some animals, such as Himalayan rabbits and Siamese cats, have mutant alleles for tyrosinase.

The enzyme is made but it is unstable and is inactivated at normal body temperature. However, the tips of their tail, paws and ears are much darker than the rest of their bodies (Figure 3.52).

Q 3.29 Suggest how the environment and genotype are interacting to produce the distinct colouring at the tail, paws and ear tips in Figure 3.52.

Q 3.30 There are rare cases in humans in which hair in the armpits is white, but hair on places such as the legs is dark. Suggest how this happens.

Figure 3.52 Only the ears, paws and tail of this Himalayan rabbit are dark.

Environment influences the epigenome

Throughout life, environmental factors can trigger changes in the epigenome and affect gene expression. These factors include medicines, drugs and other chemicals in the environment. The effect of diet on the genome has been extensively studied: experiments with mice have shown the influence of diet on early development. The mice shown in Figure 3.53 are genetically identical, but epigenetically different. The agouti gene in the normal healthy brown mouse is methylated and so not expressed. In the yellow, obese mouse the same gene is not methylated and so is expressed. The agouti protein binds to the MSH receptors in the skin and prevents the production of the dark pigment. It has also been found that the mouse brain contains similar receptors in an area of the brain that is involved in feeding behaviour: blocking these receptors may result in the mouse over-eating. However, if a pregnant, yellow mouse is fed a methyl-rich diet, her young will be slim and brown, suggesting that the epigentic change is early in development.

Figure 3.53 Genetically identical, but epigenetic changes mean the agouti gene is expressed and the phenotype altered to produce the yellow colouring. Obesity increases the mouse's risk of suffering from diabetes and cancer as an adult.

A similar effect on the agouti gene has been observed in pregnant mice whose diet included the chemical Bisphenol A (BPA), a compound used in making plastic bottles. The chemical appears to prevent the methylation of the agouti gene and, as a result, their offspring included yellow, obese mice. If the parents were fed a methyl-rich diet, the epigenetic effect of the BPA was avoided.

Behaviour can also cause epigenetic changes

Scientists at McGill university in Canada noticed that their laboratory rats varied in how good they were at being a mother to their newborn offspring (pups). 'Good mothers' were calm, and licked and groomed their pups frequently, whereas 'bad mothers' were nervous, easily frightened, and licked and groomed their pups rarely. The researchers divided their rat mothers into two groups, consisting of High LG (licking and grooming) mothers and Low LG mothers. The pups grew up to be very similar to their mothers. This suggested a genetic difference between the two groups of rats.

The scientists then swapped newborn rat pups from each group. The offspring of the High LG 'good' mothers brought up by Low LG 'bad' mothers grew up to be anxious adults. The offspring of the bad mothers brought up by good mothers were calm adults.

How can these observations be explained? It has been found that the two types of pup are epigenetically different (Figure 3.54), the GR gene in pups with low grooming and licking have been switched off due to methylation of the gene. This gene produces a receptor protein that binds the stress hormone glucocorticoid. When enough has bound, it stops the stress response by causing calming signals to be sent out, relaxing the pup after stress. (Serotonin may be involved: you will look at its role in Topic 8.) The methylation remains throughout the rat's life and means the adult rats are calmer and more likely to lick and groom their offspring. If there are low levels of the GR protein, hormone levels remain higher in the blood and the pup or adult rat is stressed for longer.

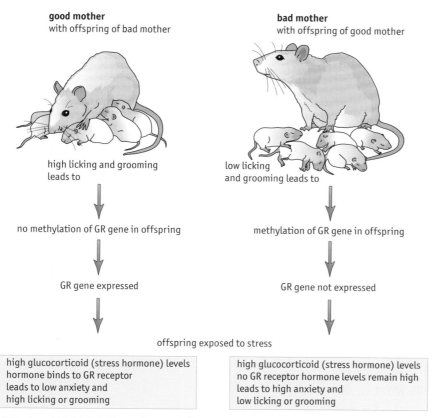

Figure 3.54 Offspring of High LG (licking and grooming) mother rats were adopted by Low LG mothers (left) and *vice versa* (right). Levels of anxiety, and licking and grooming behaviour were determined by the adoptive mother and not by the true parent.

Can the epigenetic changes be inherited?

In 2014, researchers at the London School of Hygiene and Tropical Medicine reported a study of epigenetics in 2000 mothers and babies in rural Gambia, where diets differ markedly between the dry season and the rainy season. Mothers who were pregnant in the rainy season had much better diets and higher levels of nutrients in their blood. When the infants were born they were found to have more methylation in six of their genes, suggesting that a mother's diet can affect gene expression in their unborn babies. Dr Branwen Hennig reports 'Our results represent the first demonstration in humans that a mother's nutritional well-being at the time of conception can change how her child's genes will be interpreted, with a life-long impact.'

Q 3.31 **(a)** The town of Overkalix in northern Sweden has a unique record of births, deaths and changes in food availability due to the weather going back to the 19th Century. Swedish researcher Lars Olov Bygren and colleagues found that men whose grandfathers had a season of plentiful food just before they reached puberty died several years earlier (often of diabetes) than men whose grandfathers had little food at that age. Suggest how these effects of diet might be passed on from grandparent to grandchild.

(b) The study also found that if a woman lived through regular sharp changes in food supply before puberty, her sons' daughters had a higher risk for death due to CVD.

Suggest how this effect of diet might be inherited.

Eggs and sperm are specialised cells, so it would follow that they will have changes to the epigenome that have helped determine their specialised structure and function. However, the fertilised egg divides to form stem cells that can divide to produce all types of cells. It is likely that this is achieved by the removal of the epigenetic changes during development, returning the cells to stem cell configuration. However, it is thought that some epigentic changes do pass from parent to child.

Cancer

What causes cancer?

One in three people in the UK will suffer from cancer at some stage in their life and, at present, one in four people dies from the disease. If we could understand the way our genetic make-up and our environment combine to cause cancer, we would be on the way to finding means of prevention and cure.

Cancers occur when the rate of cell multiplication is faster than the rate of cell death. This causes the growth of a **tumour**, often in tissues with a high rate of mitosis, such as the lung, bowel, gut or bone marrow. Cancers are caused by damage to DNA. DNA is easily damaged by physical factors, such as UV light or asbestos. It can also be damaged by chemicals, known as carcinogens, which may be in the environment or can be produced by cell metabolism. Mutations can also occur when cells divide (see Topic 2, Section 2.5). If DNA is copied incorrectly in gamete formation, an inherited form of cancer can result.

Research into the molecular causes of cancer has shown that epigenetic changes to the DNA and histones also have a significant role in the development of cancer. Abnormal methylation of genes in cancer cells can lead to activation or deactivation of genes that are involved in the control of the cell cycle. For example, lower levels of methylation than in healthy cells could cause activation of genes that promote cell division.

ACTIVITY

Student Activity 3.18 lets you look at how the relative contributions of genes and the environment are difficult to determine.

Q 3.32 | Explain why cancers are more likely to occur in tissues with a high rate of mitosis.

Q 3.33 | Explain why damage to the DNA in an embryo can result in inherited cancer.

Telling cells what to do

Cells go through a fixed sequence of events during the cell cycle – G_1, S, G_2 and mitosis (M) – as shown in Section 3.2. The progression from one phase to the next is controlled. During each stage of the cycle, proteins are produced that stimulate the next stage in the cycle. Cells also produce proteins that stop the cell cycle, preventing progress from one stage to the next. These proteins activate or inhibit enzymes that initiate the reactions in the next stage of the cycle.

Cancer cells do not respond to the control mechanisms. Two types of gene have a role in control of the cell cycle and play a part in triggering cancer. These are:

- oncogenes
- tumour suppressor genes.

Oncogenes code for the proteins that stimulate the transition from one stage in the cell cycle to the next. DNA mutations or epigenetic changes, such as less methylation, in these genes can lead to the cell cycle being continually active. This may cause excessive cell division, resulting in a tumour.

Tumour suppressor genes produce suppressor proteins that stop the cycle. DNA mutations or epigenetic changes – for example, more methylation inactivating these genes – mean there is no brake on the cell cycle. One example of a tumour suppressor protein is p53. This protein stops the cell cycle by inhibiting the enzymes at the G_1/S transition, preventing the cell from copying its DNA. In cancer cells, a lack of p53 means the cell cannot stop entry into the S phase. Such cells have lost the control of the cell cycle. Loss of tumour suppressor proteins has been linked to skin, colon, bladder and breast cancers.

Q 3.34 | Explain how

(a) less methylation of the oncogenes, and

(b) more methylation of the tumour suppressor genes could contribute to the development of cancer.

EXTENSION

In **Student Extension 3.7** find out how enzymes control the cell cycle at checkpoints.

There is a very complex network of signals and inhibitors that interact to control the cell cycle. There needs to be damage to more than one part of the cell control system for cancer to occur. This makes cancer very unlikely in any particular cell, but because the body contains so many cells dividing and changing over a lifetime, cancers will occasionally occur. Cancers are more likely in older people as they have accumulated more mutations.

Q 3.35 In some people, one of two alleles for the tumour suppressor protein p53 is damaged. The damaged allele is recessive to the normal allele. Explain why such people are more susceptible to environmentally induced cancer than people with two normal alleles for p53.

Inherited cancer

Most common cancers occur more frequently in close relatives of cancer patients, suggesting an inherited component. Many gene defects have been identified that predispose people to cancers including bowel cancer, ovarian cancer, prostate cancer, retinal cancer and some types of leukaemia.

For example, mutations in the gene BRCA1 predispose a person to breast cancer. The functioning BRCA1 gene produces a protein used to repair DNA. A child who inherits one defective BRCA1 allele may get cancer later in life if the other allele becomes damaged in breast tissue cells. Having a single defective BRCA1 allele does not therefore mean that breast cancer is inevitable. It simply means that such individuals are more susceptible to cancer through environmental DNA damage. Women who inherit a single BRCA1 mutation have about a 60% chance of developing breast cancer by the age of 50 compared with only a 2% chance for those who inherit two normal BRCA1 alleles. The mutation confers a high risk but is relatively rare, accounting for only 5% of breast cancer cases.

Environment and cancer

You will have noticed that newspapers and magazines are full of suggestions for living a healthy life and reducing cancer risk. How useful are these suggestions?

Damage from the environment can be either chemical or physical. The greatest chemical risk of all is from smoking. Smoking increases the likelihood of many forms of cancer, especially lung cancer, through the action of carcinogens in tar. Tar lodges in the bronchi and causes damage to DNA in the surrounding epithelial cells.

Ultraviolet light (UV) physically damages DNA in skin cells. Sometimes, a mole that has been affected by UV light may start to grow bigger and can develop into a tumour (Figure 3.55). If a tumour is not removed, cancer cells sometimes spread to other parts of the body, carried in the blood and lymphatic systems. New cancers may then form in other organs.

Figure 3.55 UV damage can cause cancer in skin cells.

Diet is also linked to cause and prevention of cancer, though the connections are not always clear. A diet with plenty of fresh fruit and vegetables provides antioxidants that destroy radicals (see Topic 1). Radicals are chemicals from the diet, from environmental factors, such as smoke and UV, or produced by the cell's own metabolism. Radicals contribute to ageing and cancer through DNA damage.

Several cancers are triggered by virus infection. For example, liver cancer can follow some types of hepatitis, and cervical cancer can follow infection by the papilloma (genital wart) virus. A virus's RNA may even contain an oncogene, which it has picked up from one of its hosts and then transfers to the cells it infects.

Q 3.36 Explain why chemotherapy and radiotherapy are often unsuccessful in tumours where the cause of the cancer is damage to the protein p53.

DID YOU KNOW?

Combating cancer

One way of treating cancer is to use surgery to physically remove the tumour. Another is to destroy the cells in the tumours. In chemotherapy powerful chemicals are used to do this and in radiotherapy X-rays or other radiation are directed at the tumour. The difficulty is to target the tumour cells without damaging nearby healthy tissues.

It used to be thought that chemotherapy and radiotherapy work by preferentially killing dividing cells in the tumour. It is now believed that these treatments actually work by inducing cells to carry out cell suicide. Chemotherapy and radiotherapy cause some DNA damage, but not enough to kill tumour cells. However, the DNA damage causes the release of the protein p53, so the cancer cells stop dividing.

ACTIVITY
Use **Student Activity 3.19** to check your notes, using the topic summary provided.

TOPIC TEST
Now that you have finished Topic 3, complete the end-of-topic test before starting Topic 4.

TREHALOSE
A SUGAR FOR DRY EYE

As we have seen in Topics 1 and 2 biological molecules have numerous roles within living organisms. As the extracts below show some functions are rather unexpected.

WHAT IS TREHALOSE?

Trehalose, as the name suggests, is a disaccharide, it is made up of two glucose molecules. It has about the same energy content as sucrose but is only half as sweet and has a lower glycemic index; less insulin is released on ingestion than with sucrose.

Trehalose is widespread in many plants and animals but it does not occur in mammals, although humans do possess the enzyme trehalase which catalyses the breakdown of trehalose.

It is thought that interaction between trehalose, water and cell structures enables organisms to survive dehydration for extended periods. The trehalose functions as an osmolyte, protecting proteins from damage by desiccation. They affect protein folding and stabilize folded proteins.

Research into the use of trehalose for treatment of neurodegenerative diseases such as Alzheimer's, Huntington's, and Parkinson's diseases has produced positive results in mouse models. These diseases result from the aggregate of misfolded proteins in the brain which leads to neural dysfunction. In trials, trehalose solution has been shown to be an effective and safe eye drop treatment for severe dry eye syndrome, protecting against corneal damage due to desiccation.

Article written by a teacher

Nishant Jumar Jain and Ipsita Roy
Protein Sci. January 2009; 18(1): 24–36.

EFFECT OF TREHALOSE ON PROTEIN STRUCTURE

Trehalose is a ubiquitous molecule that occurs in lower and higher life forms but not in mammals. Till about 40 years ago, trehalose was visualized as a storage molecule, aiding the release of glucose for carrying out cellular functions. This perception has now changed dramatically. The role of trehalose has expanded, and this molecule has now been implicated in a variety of situations. Trehalose is synthesized as a stress-responsive factor when cells are exposed to environmental stresses like heat, cold, oxidation, desiccation, and so forth. When unicellular organisms are exposed to stress, they adapt by synthesizing huge amounts of trehalose, which helps them in retaining cellular integrity. This is thought to occur by prevention of denaturation of proteins by trehalose, which would otherwise degrade under stress.

One of the most important reasons why trehalose is such an important bioprotectant is due to the existence of a number of polymorphs. The most commonly occurring crystalline form, trehalose dihydrate (depicted as Th), is stable at room temperature. Careful dehydration of the dihydrate under defined conditions leads to the formation of the anhydrous crystal.

Various theories to explain the "exceptional" properties of trehalose. (A) Vitrification theory assumes that trehalose forms a glassy matrix that acts as a cocoon and presumably physically shields the protein or indeed cells from abiotic stresses. (B) Preferential exclusion theory, on the other hand, proposes that there is no direct interaction between trehalose and protein (or biomolecule). Instead, as can be seen, addition of trehalose to bulk water sequesters water molecules away from the protein, decreasing its hydrated radius and increasing its compactness and consequent stability. (C) Water replacement theory talks of substitution of water molecules by trehalose-forming hydrogen bonds, maintaining the three-dimensional structure and stabilizing biomolecules.

Extract from review article published in the journal
Protein Science.
http://www.ncbi.nlm.nih.gov/pmc/articles/PMC2708026/

START BY REVIEWING THE SOURCE

The article 'What is Trehalose?' is unpublished, it was written for sixth form students to introduce them to a new sugar. The information is from the author's own internet research.

1. Read the articles and comment critically on the reliability of the articles as sources of scientific information.

2. Biologists sometimes express ideas in a way that implies a purpose that does not exist, this is known as teleological. It is common in ecological and environmental biological writing, for example an author may say arctic foxes produce white fur to help them survive, this implies that the fox thought about it and decided to grow white fur, it is not the case, foxes with white fur survive because they are camouflaged against the snow.

   The paragraph below appears later in the Protein Science Trehalose review article. Comment on whether you consider there is any teleology in this paragraph and justify your view.

> It is possible that depending on the structure of the protein/biomolecule in question, trehalose will be able to manipulate the water structure around itself, such that the protein/biomolecule is stabilized. Though the distribution of water molecules around trehalose is not uniform, they are oriented around trehalose in such a way that an ordered structure, with hydrogen bonds in all directions, is formed.

Biological vocabulary
As you read the articles check the meaning of unfamiliar words, e.g. glycemic index (GI) a measure of how quickly a food raises blood sugar level. Foods that cause rapid rises have a high GI.

Command word
When the word justify is used in this context you are being asked to give clear evidence to support your answer.

REVIEWING THE BIOLOGY

Having read the article, draw on your knowledge gained so far in the course and answer the following questions.

1. Suggest how trehalose in eye drop solution might prevent corneal damage that can occur in dry eye syndrome.

2. Suggest how encapsulating a protein molecule in a trehalose matrix could maintain its stability.

3. Explain why the protection of protein structure is so important for cells.

4. Suggest why humans posess the enzyme trelahose.

5. Neurodegenerative diseases such as Huntington's disease result from expansion of a CAG triplet repeat section in the gene coding for Huntingtin protein. The proteins produced aggregate together and it has been suggested that this may damage cells in the brain. Trehalose inhibits the aggregation of these proteins so may have protective role.

   a Which amino acid is coded for by the CAG triplet repeat?

   b How might an increased number of these triplet repeats affect the protein produced?

   c How might the trehalose prevent the aggregation of the proteins?

BIODIVERSITY AND NATURAL RESOURCES

Why a topic called Biodiversity and natural resources?

The Brazil nut 'story'

What do the four organisms in Figure 4.1 have in common? You are correct if you suggest that they all live in the same habitat. They can all be found living in the Amazonian rainforest. Even better if you suggest that they are dependent on each other for their survival. The story of the Brazil nut tree is amazing. The tree grows to over 50 m in height. Its fruit is the size of a grapefruit and weighs about two kilograms. Each fruit contains about 20 seeds – we know these as Brazil nuts. The fruits remain in the canopy for over 15 months before falling to the forest floor. On the ground, there is only one animal with teeth strong enough to gnaw open the extremely hard fruit – a small rodent, the agouti. Once open, the agouti eats most of the Brazil nuts immediately, but buries some to be recovered later. If the Brazil nuts remain undisturbed by the agouti or other herbivores for over a year, their shells crack and they germinate. Brazil nut seed dispersal is completely dependent on this one animal.

Pollination of the Brazil nut tree is also specialised, relying on just one type of bee. Euglossine bees (also called orchid bees) are the trees' only pollinators. They are large enough and strong enough to get into the tightly hooded flowers. Female bees do most of the pollinating, using their long specialised tongues to reach the nectar.

The male bees visit orchid flowers, which provide scent for them to attract female mates. The orchid benefits as the bee transports pollen to the next orchid flower visited. Without the orchids there would be no young bees, no adult bees and no Brazil nuts. All four species are inextricably linked: if one became extinct, the others would not be able to survive.

It is not just these species that depend on the Brazil nut tree. When it rains, any empty Brazil nut fruits on the forest floor fill with water and become home to several species. Scientists have identified a frog that breeds only in these water-filled husks. What caused these organisms to become so highly specialised, with their own continuing survival dependent on one species?

Figure 4.1 The Brazil nut tree, agouti rodent, euglossine bee and orchid are all found in the Amazonian rainforests of South America.

So many different species

The Brazil nut tree, agouti, euglossine bee and orchid are all part of a tremendously rich **biodiversity** found in rainforests. We have a multitude of different examples of organisms and species in this topic, reflecting the diversity of life on Earth. Today's biologists continue to collect, quantify and classify the living world. Yet, despite their best efforts, we have only a vague idea of how many millions of species share the planet with us.

A recent estimate suggests there are roughly 8.75 million different species on Earth. Where does this rich variety of living organisms come from and how are they so beautifully adapted to their surroundings? Charles Darwin and another naturalist, Alfred Russel Wallace, formulated a theory to answer this fundamental question, the theory of **evolution** by **natural selection**.

Biodiversity in crisis

There is growing concern about the number of species that are threatened with **extinction**. Why should we be concerned? Extinctions have occurred since life first appeared on Earth over 3500 million years ago. But the current extinction rate is many times faster than expected from the geological record and is accelerating.

Forests and other ecosystems are valuable in their own right. They are worth far more to us humans alive than when they are dead. Wild organisms provide all kinds of useful resources, such as Brazil nuts. Some are vital to our survival and wellbeing. Organisms do not do these things for our benefit. For them it is just a matter of 'survival of the fittest' and we take advantage of their ingenuity.

OVERVIEW OF THE MAIN BIOLOGICAL PRINCIPLES COVERED IN THIS TOPIC

In this topic you will study how the huge biodiversity found on Earth came about. You will first look in detail at how species are adapted to different ecological niches. You will then consider how they become adapted by means of natural selection. To understand what is meant by biodiversity, you will study how it can be classified and measured. You will also look at how classification has changed as new information is evaluated and how DNA technology is helping to improve our understanding of the evolutionary relationships between living organisms.

Building on what you have studied in earlier topics, you will compare the structure and function of starch and cellulose, and contrast plant and animal cell structure. You will examine specialisation of cells for support and transport through plant stems.

You will consider the value of biodiversity in providing natural resources, focusing on the uses we make of plants. You will see that humans have relied on plants throughout history and that the use of plant resources can contribute to sustainable living.

Finally, you will examine the role of zoos and seed banks in the conservation of endangered species.

ACTIVITY

Darwin discovered numerous new species – many of them in the Galapagos Islands, lying 1000 km off the west coast of South America. In **Student Activity 4.1** you can take a tour of the Galapagos Islands to see the variety of life that Darwin encountered.

ACTIVITY

Student Activity 4.2 lets you develop some of the observation and interpretation skills that Darwin used and that today's biologists still use.

REVIEW

Are you ready to tackle Topic 4 Biodiversity and natural resources? Complete the GCSE review and GCSE review test before you start.

4.1 Why are there so many different species?

Every available habitat on Earth is colonised by living organisms, from frozen ice fields on high mountains to tropical rainforests. Over 1.58 million **species** have already been recorded in the online Catalogue of Life (Figure 4.2).

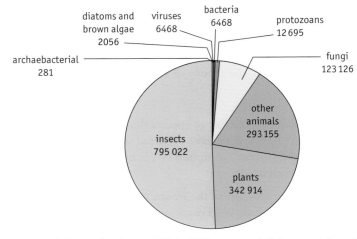

Figure 4.2 Species totals recorded in the Catalogue of Life in 2014: see page 163 for more information about this online catalogue. It is thought that 86% of species on land and 91% of species in the ocean are yet to be discovered.

WEBLINK

Visit the Catalogue of Life website to find out the latest figures for number of species recorded.

KEY BIOLOGICAL PRINCIPLE: WHAT IS A SPECIES?

What is a species? Perhaps surprisingly, this is one of the most difficult questions that biologists try to answer. As you can see from Figure 4.3, it is not enough to say that a species is a group of individuals that closely resemble one another. The most widely used definition is:

A species is a group of organisms with similar morphology, physiology and behaviour, which can interbreed to produce fertile offspring, and which are reproductively isolated (in place, time or behaviour) from other species.

Figure 4.3 This is not a mouse. It is a marsupial and is more closely related to kangaroos and koalas than to rodents like mice or agoutis. Similar-looking organisms are not necessarily closely related.

Although horses and donkeys have, to some extent, similar morphologies (features), physiologies and behaviours, and can be bred together to produce mules, the mules are infertile. This means that, although horses and donkeys are closely related, they remain distinct species. However, some species do not seem to know this rule, and will quite happily breed and produce hybrid offspring that are fertile. The thirteen finch species of the Galapagos Islands have been known to interbreed and produce viable offspring. If this happened frequently we would expect small islands such as Daphne, with only four species of finch, to end up with birds of average types and no distinct species. But the finch species do remain distinct, because the hybrid forms do not survive drought conditions as well as the pure-bred types. There is enough fluctuation in the island's climate to maintain the different species.

Biologists sometimes disagree about a particular species and sometimes one species is reclassified into two. Marsh and willow tits were once thought to be the same species (Figure 4.4). In 1999 the common pipistrelle bat was split into two species on the basis of differences in their echolocation calls. The common pipistrelle (*Pipistrellus pipistrellus*) calls at a frequency of 45 kHz, while the soprano pipistrelle (*P. pygmaeus*) calls at 55 kHz.

Sometimes two separate species are reclassified as a single species. For example, the locust species *Locusta migratoria* and *Pachytylus migratoria* were merged into one species, known as the migratory locust (*Locusta migratoria*).

Figure 4.4 The willow tit (left, *Parus montanus*) and marsh tit (right, *Parus palustris*) were thought to be just one species until about 1900.

Modern techniques of DNA analysis are increasingly important in the identification of species. DNA analysis can help to tell us whether two slightly different populations are one species or two. For example, the British northern brown argus butterfly was thought to be a different species from Scandinavian populations, but analysis of mitochondrial DNA shows that they are the same species, *Aricia artaxerxes* (Figure 4.5).

DNA 'bar codes' are being developed to identify species without the need to look at their phenotypes. These work rather like the DNA fingerprints used in forensics, which you will study in Topic 6. In the future it may be possible to extract DNA from a plant in the field and identify it on the spot by comparison with a database, using a satellite link.

However, DNA cannot always give a clear answer about species, because it cannot tell us whether two populations can breed together to produce fertile offspring. We have to observe the two living populations in natural conditions in order to find out if they can do this.

The decision about whether or not a population is one species or two is ultimately a matter of judgement. As Darwin found, isolated populations, such as those found on islands, often show differences from those on the mainland.

Figure 4.5 A brown argus butterfly, *Aricia artaxerxes*.

 Q 4.1 Read the descriptions below and decide if the groups are the same species or different species. Explain your answers.

(a) Group A has 98% of its genes in common with group B.

(b) Group C can interbreed with group A and produce fertile offspring.

(c) Group D is physically very similar to group B, except that it is a different colour. Where groups B and D overlap in geographical areas, mating between them produces offspring of an intermediate colour. The males of these offspring are infertile.

(d) Group E looks very similar to group F, but has darker feathers. The genetic fingerprints of groups E and F are identical.

Species occupy different niches

Habitats and communities

Take a walk through a rainforest or a local wildlife area and you will soon see that the area is not uniform. There can be many different **habitats**. A habitat can be thought of literally as the place where an organism lives. A moorland, for example, is made up of undulating heather-clad slopes, streams, bogs and rocky outcrops. Each habitat has a particular set of conditions that supports a distinctive combination of organisms.

Within a habitat there may be many populations of organisms. Each **population** is a group of interbreeding individuals of the same species found in an area. The various populations in a habitat make up a **community**. So, for example, a rainforest community consists of all the organisms that live in the rainforest, which might include the Brazil nut trees, agoutis, euglossine bees, frogs and orchids.

An organism's environment provides all its essential resources, such as energy, raw materials, a place to live and a mate. Most resources needed for survival are in limited supply. In order to survive and reproduce, organisms must compete successfully for the resources they need, against other organisms of the same and different species.

Different niches avoid competition

Two species sharing the same habitat tend not to be in competition with each other. Each species occupies a particular **niche**, often referred to more fully as an ecological niche. A niche can be defined as 'the way an organism exploits (uses) its environment'. All the species sharing a habitat have different niches. If two species live in the same habitat and have exactly the same role within the habitat – the same food source, the same time of feeding, the same shelter site and so on – they occupy the same niche and they will compete directly with each other. The better-adapted organisms will out-compete the other and exclude it from the habitat. This happened to the red squirrel when the North American grey squirrel was introduced to England in the nineteenth century. The native red squirrel has been out-competed and replaced by the grey squirrel in much of England.

Orchid niches

The niches of some orchids exploit the behaviour of insects as part of their highly specialised pollinating mechanisms. As you saw in the introduction to this topic, female euglossine bees visit the flowers of Brazil nut trees to collect nectar. At the same time, they inadvertently pick up and carry pollen from one tree to the next, pollinating the flowers. The male euglossine bees visit orchids to feed and to collect aroma chemicals, which are important in attracting a female mate.

Pyramidal orchids, a common sight on chalk and limestone grassland across England and Wales, produce scent that attracts day-flying moths. As a moth probes into the deep, tapering corolla of the orchid flower, a parcel of pollen in the shape of a collar attaches itself with quick-drying glue around the moth's tongue. After a short time, the sides of the pollen mass curl forward. The pollen is now in just the right position to pollinate the two stigmas of the next flower the moth visits.

Some orchids are so specialised that they can only be pollinated by one species of insect. For example, the mirror orchid (*Ophrys speculum*) is pollinated by just one species of wasp. Only male wasps visit the orchid, which resembles a female wasp of that species. The male mistakes the flower for a female wasp and, as he attempts to mate, the pollen mass sticks to his head. He then carries this to the next flower as he attempts to mate again. The flower even produces a scent similar to the pheromone made by the female wasp to attract males. About 400 species of orchid worldwide are known to mimic female insects and their pollinators are tricked into 'pseudocopulation' (Figure 4.6). By doing so, they out-compete other flowering plants for the attentions of insect pollinators.

Q 4.2 Suggest (a) an advantage and (b) a disadvantage of using just one type of insect pollinator.

Woodpecker niches

There are three species of woodpecker in Britain: the lesser spotted, the greater spotted and the green woodpecker. All woodpeckers have a powerful beak that they use for probing into rotting wood. They pick up food such as woodlice, beetle larvae and ants using a very long tongue. The tiny lesser spotted woodpecker prefers the finer branches at the tops of trees, whereas the greater spotted tends to feed on the broader branches (Figure 4.7). The green woodpecker tends to feed on the ground. All three species can co-exist in the same habitat because each has a different niche.

Woodpeckers occur throughout most of the world, but they are not good at flying over large stretches of ocean and so have never reached Australasia, Madagascar or the Galapagos Islands. In Madagascar, the woodpecker niche is occupied by the aye-aye, which uses its long bony middle finger in a similar manner to a woodpecker's beak. Aye-ayes are **endemic** to Madagascar, which means that is the only place they are found. In the Galapagos Islands, off the coast of Ecuador, no species has yet evolved a comparable structure to the woodpecker's beak. However, one small bird, called the woodpecker finch, has acquired the habit of breaking off a cactus spine and using it to probe into rotting wood, just like a woodpecker uses its tongue.

Q 4.3 What do you think might happen to the woodpecker finch if a pair of breeding woodpeckers arrived in the Galapagos Islands?

Figure 4.6 Digger wasp 'mating' with a fly orchid flower. When the female wasps emerge two weeks after the males, the flowers are no longer visited.

⚙ ACTIVITY

In **Student Activity 4.3** you can work out how one bee is exploiting its niche.

Figure 4.7 Different feeding strategies mean that this greater spotted woodpecker does not compete with other woodpecker species.

Adapted to their environments

Woodpeckers, aye-ayes, orchids and all other organisms are **adapted** in such a way that they are uniquely able to exploit their own particular niches. Being adapted means being specialised to suit the environment in which the organism lives. Features that enable organisms to survive are called **adaptations**. Adaptations can be classified as behavioural, physiological or anatomical, though these are not hard and fast categories – there is some overlap.

Behavioural adaptations

Behavioural adaptations are any actions by organisms that help them to survive or reproduce. When agoutis bury Brazil nuts, this is a behaviour that will aid their survival by providing a future food source. Even plants respond to their environments, for example, as they turn their leaves towards the Sun, maximising the amount of light received for photosynthesis. The explosive dispersal of plant seeds, for example, in balsam and bitter cress, could also be described as a behavioural adaptation. Some fungi use explosive mechanisms to expel their spores. The Costa Rica palm, *Socratea durissima*, has stilt-like roots that grow towards a sunny patch and slowly pull the whole tree towards it.

Q 4.4 For each of the behaviours described below, suggest what the organism is responding to and how the behaviour helps it survive.

(a) Herring gull chicks peck at the red spot on the parent's beak, causing the parent to regurgitate food.

(b) Some ground-nesting birds, such as ringed plovers, perform a 'broken wing display' on seeing a predator (Figure 4.8). The parent bird runs away from the nest, trailing its wing as if it is broken. But just as the predator is approaching, the bird takes off and escapes.

(c) In drier conditions, woodlice move around more. In humid conditions they move about much less.

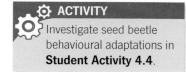

ACTIVITY
Investigate seed beetle behavioural adaptations in **Student Activity 4.4**.

Figure 4.8 Broken wing display by ringed plover.

Physiological adaptations

Physiological adaptations are features of the internal workings of organisms that help them to survive or reproduce.

British motorists in springtime may notice a swathe of tiny white flowers along the very edges of the grass verge. The flowers are Danish scurvy grass, *Cochlearia danica,* a member of the cabbage family. It was formerly found mainly by the sea, but has spread inland with the increasing use of salt on icy roads (Figure 4.9). The plant has a physiological adaptation allowing it to tolerate high salt concentrations. It has therefore been able to occupy a newly created niche unavailable to other inland plants.

A

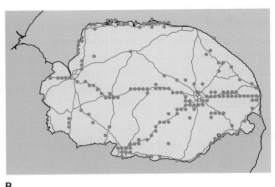
B

Figure 4.9 A Danish scurvy grass, *Cochlearia danica,* used to be confined to coastal habitats. **B** The distribution of Danish scurvy grass in Norfolk. Red lines show main roads.

Q 4.5 Humans are unable to survive by drinking sea water, but whales, fish and many marine birds are able to take in sea water and excrete the excess salt. Salt excretion is achieved by ion pumps in kidneys, fish gills and in the beaks of some birds. Name the process by which salts are pumped across membranes using energy from respiration supplied by the energy transfer molecule ATP.

Hot springs make a very extreme environment, yet some organisms are adapted to them physiologically. Thermophilic bacteria can tolerate the highest temperatures of any organism. Those found in thermal vents deep in the ocean can survive at 350 °C. Others live in hot springs, such as those at Yellowstone National Park, at temperatures up to 80 °C. They achieve their heat tolerance by having heat-stable enzymes. Such enzymes have useful applications for humans, such as the production of biological washing powders.

Anatomical adaptations

Anatomical adaptations are the structures we can see when we observe or dissect an organism. For example, the bodies of bumblebees show adaptations used to collect nectar and pollen. There are about 250 species of bumblebee worldwide. Each has a long tongue through which it can suck nectar (sugary solution) from flowers (Figure 4.10). Within a habitat, some bumblebee species have longer tongues than others, so they are specialised to feed from different flowers.

Q 4.6 Table 4.1 shows the flowers visited by six species of bumblebee seeking nectar in an English garden. The flowers are listed in order of increasing corolla length (length of petal tube) and the bumblebees in order of increasing tongue length. Explain what the data show about the niches of these six bumblebee species, when competing for nectar in this garden.

Bumblebees also have a pollen basket on their hind legs. This is a fringe of hairs that holds pollen combed from the body (Figure 4.10), allowing pollen to be carried back to the nest where it will be used to provide food for the larvae. A slot on the front legs is used by a bumblebee to clean any pollen grains stuck on its antennae.

⚙ ACTIVITY

In **Student Activity 4.5** discover more animal and plant adaptations and listen to Sarah Darwin talking about the Galapagos tomato. Endemic to the islands, it is a salt-resistant plant able to grow in salty soil.

Figure 4.10 Bumblebee (Queen *Bombus lucorum*) with full pollen baskets.

	Corolla length/mm	Bombus pratorum	Bombus lapidarius	Bombus lucorum	Bombus terrestris	Bombus pascuorum	Bombus hortorum
Tongue length/mm		5	6	7	8	13	15
Mallow	0	V					
Daisy	0	V	V				
Poppy	0		V				
Yellow clover	1		V				
Shepherd's purse	2		V				
Bramble	3	V		V	V	V	V
White clover	3	V	V	V			V
Bird's foot trefoil	4		V				
Snowberry	4	V				V	V
Gooseberry	4	V					
Broccoli	5	V				V	V
Woundwort	8					V	V
Delphinium	8						V
Foxglove	10					V	V
White dead-nettle	11					V	V
Sage	12					V	V
Sweet William	15						V
Broad bean	15						V
Honeysuckle	35						V

Table 4.1 Flowers visited by six species of bumblebee (*Bombus*) with increasing tongue length. V = visited.

Q 4.7 Describe one adaptation in each of the species in the Brazil nut 'story' at the start of the topic and in each case state if it is a behavioural, physiological or anatomical adaptation.

As you saw in the Brazil nut and orchid bee example, plants and insects can evolve in tandem. The plant and its pollinator become dependent on each other, and more and more closely adapted. This is called **co–adaptation**.

Another example of co-adaptation is the whistling thorn, a species of acacia tree found on African savannahs. The tree has pairs of thorns on its stems. At the base of some of the thorns are swellings (Figure 4.11). Stinging ants make their home inside the swellings and swarm out when browsing animals, such as giraffes, try to eat the tree's leaves. In this way the ants deter the herbivores.

Q 4.8 Suggest how the ants benefit from living in the swollen spines of the whistling thorn described above.

Figure 4.11 The swollen spines of the whistling thorn, *Acacia drepanolobium*, also known as the swollen spine thorn. When the swollen spines are old and no longer inhabited by ants, wind blows through the holes made by the ants and the resulting whistling gives the tree its name.

CHECKPOINT

4.1 The structures of roots, leaves, stems, spines, flowers and fruits all provide beautiful examples of anatomical adaptations in plants. Describe one way that each is adapted and suggest how the adaptation aids the plant's survival.

4.2 How did organisms become so well adapted?

Natural selection

Brazil nut trees, agoutis, orchids and all other species are adapted in such a way that each is uniquely able to exploit its own ecological niche. As a population increases in size, a greater proportion of individuals will die or fail to reproduce owing to competition for resources, such as food and space. Disease and extreme environmental conditions may also cause the deaths of some individuals. This striving for survival is known as the 'struggle for existence'.

In this struggle for existence there will be winners and losers. Winners are those surviving individuals who, by chance, possessed some characteristic that gave them an advantage over others. This differential survival is **natural selection**, sometimes called 'the survival of the fittest'. (Here, 'fit' means well adapted, rather than physically fit in an athletic sense.) In the case of orchid bees it might be the possession of a slightly longer tongue, which improves their ability to compete for food. They have a selective advantage over other orchid bees, and are more likely to survive and reproduce, passing on the alleles for a longer tongue to their offspring. Individuals with longer tongues will be more common in the next generation. The average tongue length of the next generation of bees will be slightly greater. Such a change in form (or behaviour or physiology) over generations is called **evolution** by natural selection.

Natural selection is the mechanism by which species change over time. This theory of evolution by natural selection was first proposed by Darwin and Wallace, and detailed in Darwin's On the Origin of Species published in 1859.

ACTIVITY

Student Activity 4.6 lets you investigate natural selection.

Evolution by natural selection

Evolution is more precisely defined as 'a change in allele frequency in a population over time (generations)'. In our orchid bee example, there has been a change (increase) in the relative frequency of alleles for a long tongue. The emphasis on alleles in this definition reminds us that for natural selection to lead to evolution, there must be some genetic variation in the population. An allele can be selectively neutral (i.e. has no advantage or disadvantage), but suddenly become selectively very advantageous when the environment changes. The allele frequency will then increase over time.

Q 4.9 Would you expect bee tongues to go on getting longer and longer with each generation as a result of natural selection? Give a reason for your answer.

The fundamentals of the theory are summarised in Table 4.2.

1	A population has some naturally-occurring genetic variation with new alleles created through mutations.
2	A change in the environment causes a change in the selection pressures acting on the population.
3	An allele that was previously of no particular advantage now becomes favourable.
4	Organisms with the allele are more likely to survive, reproduce and so produce offspring.
5	Their offspring are more likely to have the allele, so it becomes more common in the population.

Table 4.2 Evolution by natural selection.

Q 4.10 The pictures in Figure 4.12 on page 159 show the events in a hypothetical instance of evolution. Match the statements in Table 4.2 with the correct pictures.

Figure 4.12 Evolution in action.

Natural selection is happening right now

Natural selection is occurring all the time, but it can be very slow and difficult to observe during a human lifetime. If the environment remains stable, then organisms may simply become better and better adapted to their existing niches, as Figure 4.13 shows. But if the environment changes, then selection changes too. For example, our use of pesticides and antibiotics has led to the evolution of resistant forms of insects and bacteria. One example you may be familiar with is evolution within head lice (Figure 4.14).

About 7% of UK primary school children have an infestation of head lice in their hair. These parasites suck blood and have a rapid rate of reproduction, doubling in number every seven days. There are many head lice shampoos that contain insecticides. Permethrin and malathion are frequently the active ingredients. However, in recent years there has been both anecdotal and

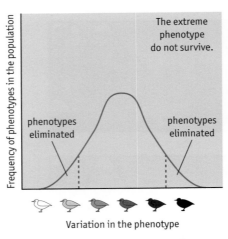

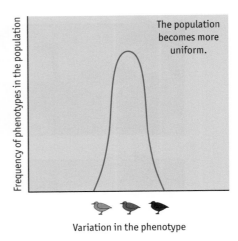

In this example the lightest and darkest phenotypes are more conspicuous to predators

Figure 4.13 Stabilising selection.

Figure 4.14 Coloured scanning electron micrograph (SEM) of a human head louse, *Pediculus humanus-capitis*. An egg (nit) has been laid on the hair (lower right). The lids of the nits (top) are perforated with air holes that allow the developing nymphs to breathe. Adult female lice lay between 80 and 100 eggs in a lifetime.

✓ **CHECKPOINT**

4.2 Identify three key steps in the process that resulted in resistant head lice.

scientific evidence to show that head lice have become resistant to these shampoos and that the shampoos are no longer as effective. Results of a study carried out at the University of Bristol showed that there was an 87% failure rate for permethrin and a 64% failure rate for malathion in the treatment of infested school children.

This is a modern example of natural selection in action. Initially, many head lice must have been killed by the insecticides, but now it seems that permethrin will kill only 13% of them. Most head lice survive and quickly breed the next resistant generation.

In this example, the insecticides in the shampoos provide the selection pressure. Had there not been pre-existing variation in the population, all of the lice on treated heads would have died. Without those original chance mutations the head louse could have become extinct – at least in areas where the chemical treatments against it were used.

Evolution has occurred: the insecticide-resistance alleles in head lice went from being quite rare in pre-insecticide populations to being extremely common.

In the case of the resistant head lice, evolution by natural selection has occurred, producing a population that has different allele frequencies from the original population. The frequency of alleles in the gene pool has changed.

Gene pools

A **gene pool** consists of all the alleles of all the genes present in a population.

Imagine all the people in your school or college taking out a set of their genes and throwing them into one big box; 20 000 or so genes from each person. The box would then contain the gene pool of the school/college population. In the gene pool box there will be the different alleles of each gene thrown in. Some alleles of each gene will be common in the gene pool and others will be rare.

The concept of a gene pool is useful when thinking about the biodiversity and adaptability of any species. Populations with a bigger gene pool – more different alleles of each gene – are said to have greater genetic diversity. They are more likely to posess alleles that will allow them to survive. The frequency of those alleles will change over time due to natural selection.

Working out allele frequencies

It is possible to calculate allele frequencies if we make the assumption that for a gene with only two alleles, such as straight thumbs (T) and curved thumbs (t), the frequency of the dominant allele, T, in a population's gene pool is p and the frequency of the recessive allele, t, in the gene pool is q. The relationship between the two frequencies is $p + q = 1$ because the only outcome for each allele is T or t. Unfortunately, you cannot directly observe allele frequencies in populations of diploid organisms. However, you can observe the phenotypes of the individuals and then use these to work out the frequencies of the alleles in the population.

In 1908, two different scientists, Hardy, an English mathematician, and Weinberg, a German physicist, each independently came up with the equation that relates phenotype frequencies that can be observed to allele frequencies. The Punnett square in Figure 4.15 shows the basis of the relationship. Their equation is called the **Hardy–Weinberg Equation** and is:

$$\underset{p^2}{\text{frequency of homozygous dominant individuals}} + \underset{2\,pq}{\text{frequency of heterozygous individuals}} + \underset{q^2}{\text{frequency of homozygous recessive individuals}} = 1$$

If we know the frequency of a phenotype in a population, usually the recessive phenotype, we can use the Hardy–Weinberg equation to calculate the frequency of each allele and the three genotypes. For example, in a population of 100 people, 91 have straight thumbs and nine have curved thumbs. The 9% with curved thumbs are homozygous recessive, tt. Using the Hardy–Weinberg equation to calculate the frequency of the alleles, it is best to change the percentage to a proportion: 9% is 0.09.

Therefore, frequency of homozygous recessive individuals, q^2, is 0.09.

so $q = \sqrt{0.09}$
     $= 0.3$

Because $p + q = 1$
then $p + 0.3 = 1$
therefore $p = 1 - 0.3$
               $= 0.7$

We have worked out that the frequency of q, the recessive allele, is 0.3 or 30%; and that the frequency of p, the dominant allele, is 0.7 or 70%.

We can also work out the frequency of the genotypes, which will be distributed according to the Hardy–Weinberg equation, $p^2 + 2pq + q^2 = 1$. The frequency of the homozygous dominant individual, TT, is p^2: $0.7^2 = 0.49$ or 49%. The proportion of the population with the heterozygous genotype, Tt, is $2pq$: $2 \times 0.7 \times 0.3 = 0.42$ or 42%. The frequencies should add up to 1 or 100%. This is the case in our example: $0.49 + 0.42 + 0.09 = 1$.

The Hardy–Weinberg equation can be used to see if there has been a change in allele frequency over time.

Q 4.11 Use the Hardy–Weinberg equation to calculate the frequency of heterozygous individuals in a population of cats in which 57% have long hair that makes them homozygous recessive for the shorthair gene: their genotype is ss.

Q 4.12 Phenylketonuria (PKU) is a genetic condition that affects how the body breaks down protein, and can result in impaired brain development if left untreated. 1 in 10 000 of the UK population have the condition; they are homozygous recessive for the mutation in the PKU gene that causes the condition. Use the Hardy–Weinberg equation to work out the percentage of people in the UK who will be carriers of this condition.

Q 4.13 (a) A Year 12 biology class tests the ability of every one in their year to taste the bitter chemical PTC (phenylthiocarbamide). Tasting is due to a dominant allele, whereas not tasting is due to the recessive allele. They found that of the 243 students, 81 could not taste PTC. Work out the frequency of the recessive allele in this population of students.

(b) The experiment was repeated five years later, this time 65 of the 215 Year 12 students tested could not taste. Has the frequency of the recessive allele changed, give some evidence to support your answer.

Gametes Frequency	T p	t q
T p	TT p^2	Tt pq
t q	Tt pq	tt q^2

Figure 4.15 When gametes form, a proportion of them, p, will contain the dominant allele (T) and a proportion of them, q, will contain the recessive allele (t). When they combine at fertilisation, the frequency of the genotype TT is p^2, Tt is $2pq$, and tt is q^2.

ACTIVITY
You can calculate allele frequencies using the Hardy–Weinberg equation in **Student Activity 4.7**.

Being adaptable

The ability of a population to adapt to new conditions will depend on:

● the strength of the selection pressure

● the size of the gene pool

● the reproductive rate of the organism.

Insects, such as mosquitoes, have rapidly become resistant to insecticides, such as DDT, because (i) the selection pressure is strong (DDT originally killed more than 99% of all mosquitoes), (ii) there are millions of mosquitoes with a very large gene pool and (iii) mosquitoes reproduce very quickly, so any surviving (resistant) ones can quickly build up numbers, passing on their resistance to their offspring.

Q 4.14 **(a)** What might cause a small number of mosquitoes to be naturally resistant to an insecticide?

(b) Explain why it is important for the survival of large mammals, such as rhinos and whales, to conserve a large and varied gene pool.

The current rate of climate change due to global warming is an example of selection pressure. Some organisms will survive and some will not. Topic 5 in the A2 course discusses the implications of global warming for species that are and are not adaptable to it.

Being perfect

Can populations become perfectly adapted to their environments? The answer is no, for the following reasons. When the environment changes, there is a time lag before a population can adapt by natural selection, so organisms are always playing a game of evolutionary catch-up. Even in a stable environment, mutations are occurring all the time, many of which are harmful to the organism. This makes a population less well adapted to its surroundings.

Being perfect, or close to perfect, has its benefits and drawbacks. Becoming adapted to a specialised niche helps avoid competition with other species living in the same area. The specialisation of the species involved in the Brazil nut 'story' means they avoid competition for food, breeding ground or other features of their niche. However, their 'perfection' also makes them interdependent and vulnerable: each species would be severely affected if any one of the others became rare. On the other hand, species that can adapt to a wide range of habitats will always be in competition with a range of other species, but they are less vulnerable to changes in their habitat.

Formation of new species

Darwin thought that, over time, populations exposed to selection pressure would become so different from their original form that they would eventually become different species. It is now generally accepted that, for a new species to arise, there has to be **reproductive isolation** of a group of individuals from the rest of the population, with each accumulating different allele frequencies.

The most common method of reproductive isolation is when part of a population is isolated by some geographical feature such as a high mountain range, a river, or a stretch of ocean. This geographic isolation prevents a group of individuals from breeding with the rest of the population (Figure 4.16). Over time, the two groups will become less like each other as they respond to different selection pressures, and as random mutations accumulate. Eventually members of the two groups may not be able to interbreed if the differences between them are great enough. Once the two populations are unable to breed and produce fertile offspring, they are considered to belong to two different species. The formation of a new species is called **speciation**. The longer the two groups are isolated, the more allele frequencies change and the more likely it is that speciation will occur.

There are a number of other ways that two populations can be reproductively isolated, for example they may occur in the same area but differences in timing of mating prevents them from breeding. In Topic 5 you will look in more detail at the variety of mechanisms that lead to reproductive isolation.

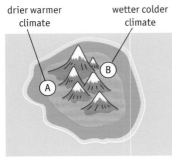

drier warmer climate   wetter colder climate

Isolated from population A, a new species, B, evolves due to the different selection pressures on that side of the mountain range.

Figure 4.16 Illustration of how geographical isolation can lead to the evolution of new species.

4.3 Quantifying biodiversity

What is biodiversity?

'Biodiversity' is a shortened form of the words 'biological diversity'. Biodiversity is used in everyday language to mean the variety of life, and in particular the wealth, of different species that exists as a result of evolution by natural selection. In biological terms it refers to the variety of species that belong to every different group of organisms – animals, plants, fungi, bacteria and other microorganisms – living in all the habitats on the planet. Biodiversity also refers to the diversity within species.

> **⚙ ACTIVITY**
>
> In **Student Activity 4.8** you can consider different ways of defining biodiversity.
>
> **Student Activity 4.9** lets you read about biologists involved in biodiversity research and helps you investigate the rate at which new species are discovered.

Current estimates of the actual number of species on Earth range from 3 million to 30 million or more. A recent estimate based on predictions for each taxonomic group suggested 8.75 million with a possible error of ±1.3 million. In just one animal group alone, the beetles, there are thought to be at least 1 million species. So far, only about 236 000 beetle species have been recorded in the Catalogue of Life, perhaps only a quarter of the world's beetle diversity. The situation is probably far worse in other less well studied or less popular groups, such as bacteria and fungi, where only a tiny fraction of the species have been given a name. We still have no rigorous estimate of how many species we share the planet with. This is not because of any particular technical difficulties. It is simply because the size of the task is massive.

Biologists want to be able to quantify biodiversity. Once biodiversity has been assessed, the data can be mapped on both a local and a global scale to reveal patterns in diversity. This can help to focus conservation efforts on vulnerable habitats or species. However, to do this and to understand the Earth's biodiversity, biologists need to define what is meant by biodiversity and be able to measure it. This is only possible if they can identify and classify the species observed.

> **⚙ EXTENSION**
>
> In **Student Extension 4.1** you can read more about where estimates of species numbers come from.

> **WEBLINK**
>
> Visit the Catalogue of Life website to explore their encyclopedia of life and find out why global species databases are so important.

> **DID YOU KNOW?**
>
> ### Producing a species catalogue
>
> The international Species 2000 Programme was set up in September 1994 with the objective of making a catalogue of all the known species of plants, animals, fungi and microbes on Earth as the baseline dataset for studies of global biodiversity. By 2000, its 'Catalogue of Life' contained 220 000 species. This had risen to over 1.5 million by 2014, which is a sizeable proportion of the estimated 1.75 million species that have been described and named. The programme was established by the International Union of Biological Sciences (IUBS) and is based at the Naturalis Biodiversity Centre in Leiden in the Netherlands.
>
> The efforts of the Species 2000 Programme may seem impressive compared with the mere 12 000 or so species described by Carolus Linnaeus. Linnaeus, a great Swedish biologist, made the first attempt to produce such a global species catalogue (*Systema Naturae*, published in 1758). However, the task still facing the Species 2000 Programme remains huge. This means that a great deal more exploration, discovery and naming has still to be done by biologists.
>
> New technology will help speed up the process. Geographical positioning systems (GPSs) can quickly pin-point the location of new species, whilst the Internet can be used to store and exchange data from around the world. Crucially, DNA analysis can now help to confirm that 'new' species are indeed different from existing species.

Unlocking identities

Putting a name to the species they find represents a significant challenge to biologists working in the field. Whether they are a research worker studying agoutis in a tropical rainforest, or a bird watcher looking at wildfowl on an estuary in the UK, accurate biological identification is vital.

Unique names

All organisms are given a scientific name. This avoids the confusion that can arise when common names are used. For example, *Arum maculatum* (Figure 4.17) has more common names than any other plant in Britain, including lords and ladies, and cuckoo pint. The Brazil nut is known locally as the castana: its scientific name is *Bertholletia excelsa*. Probably the most important thing that Carolus Linnaeus, the Swedish biologist, did in his early attempts to catalogue living things was to come up with a system in which each species was given a unique two-part Latin name. This **binomial system** is still in use today. The first part of the name, the **genus**, is shared by all closely related species; so all horses and zebras are in the genus called *Equus*. The second part of the name defines the particular species in the genus. Together, these two words make up a unique species name that is often highly descriptive (Figure 4.18).

Figure 4.17 More than a dozen common names have been recorded in Britain for *Arum maculatum*. Species names are always written in italics or underlined. The first part of the name, the genus, always starts with a capital letter. The second part of the name does not.

Figure 4.18 This bat has the scientific name *Rhinolophus ferrumequinum*, from *rhinos* (Greek for nose), *lophos* (Greek for crest), *ferrum* (Latin for iron) and equinum (Latin for horse), providing an excellent description of its nose – if you understand Greek and Latin. Its common name is the greater horseshoe-nosed bat.

Q 4.15 Using the common and scientific names below, **a** state which genus creeping buttercup belongs to and **b** decide whether meadow sweet and dropwort, or lady's mantle and dropwort, share more similarities, and give a reason for your answer.

Common name	Scientific name
creeping buttercup	*Ranunculus repens*
meadow sweet	*Filipendula ulmaria*
lady's mantle	*Alchemilla vulgaris*
dropwort	*Filipendula vulgaris*

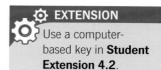

⚙ **EXTENSION**
Use a computer-based key in **Student Extension 4.2**.

DID YOU KNOW?

Biological identification

To name an organism you have to be able to identify what it is. Keys are a traditional means of identifying organisms. You will no doubt be familiar with the use of a simple text-based **dichotomous key**, which means that there are always two alternatives at each stage in the key. Versions of this type of key are routinely used by field biologists to find the names of unfamiliar organisms.

Traditional dichotomous keys have to be used in a set sequence, starting with the first pair of statements. Sometimes this can be a problem because if a characteristic referred to early in the key is absent due to damage or the stage of development of the organism, it is difficult to continue to use the key. Multiple access keys are more flexible. They allow the user to start with whichever characteristic they wish and work through the key in a different order.

Often regarded as one of the most traditional areas of biology, identification methods are increasingly making use of new technology. The field of computer-assisted taxonomy (CAT), which uses computer-based keys and electronic techniques, is expanding and beginning to replace paper-based keys. Computer-based keys or apps can allow multiple access and can easily be updated when new species are discovered.

Q 4.16 Use the key in Figure 4.19 to identify the butterfly fish shown in Figure 4.20.

Decision point	Choices	Decision
1	Pelvic fin dark	2
	Pelvic fin light	4
2	Two large white spots below dorsal fin	C. quadrimaculatus
	Lacks two large white spots below dorsal fin	3
3	Tail with two dark bars at tip	C. reticulatus
	Tail with one dark bar at tip	C. kleinii
4	Posterior or dorsal fin has long filament extension	5
	Filament extension lacking from dorsal fin	6
5	Large dark spot on body near filament	C. ephippium
	Small dark spot on body near filament	C. auriga
6	No vertical band through eye	C. fremblii
	Vertical band through eye	7
7	Incomplete eyeband on face (does not go to top of head)	C. multicinctus
	Complete eyeband on face (extends to top of head)	8
8	Nose area with band	9
	Nose area lacks band	10
9	Fewer than eight diagonal bands on body	C. ornatissimus
	More than eight diagonal bands on body	C. trifasciatus
10	Distinct white spots split eyeband above eye	C. lineolatus
	No white spot above eye; eyeband not split	11
11	Upper third of body under dorsal fin dark	C. tinkeri
	Upper third of body under dorsal fin not dark	12
12	Distinct small spots arranged in rows	13
	No distinct small spots; body has large spot or band	14
13	No black band on caudual fin	C. citrinellus
	Obvious black band on caudal peduncle	C. miliaris
14	Side with a large black teardrop; no dark bars on tail	C. unimaculatus
	Large black shoulder patch; tail with dark bars	C. lunula

Figure 4.19 Dichotomous key for butterfly fish of the genus *Chaetodon*.

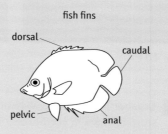

fish fins

dorsal, caudal, pelvic, anal

Figure 4.20 Use the key in Figure 4.19 to identify this butterfly fish and answer question 4.16.

Sorting and grouping

A hierarchical system

Faced with such a bewildering variety of living things, humans have always tried to organise and make sense of the variety of life. Placing organisms into groups based on shared features, known as classification or **taxonomy**, results in a manageable number of categories and has been the principal aim of all classification systems. Linnaeus created the first classification system, grouping organisms according to their visible similarities and differences.

On coral reefs there is often an amazing array of colourful fish. But on closer inspection, similarities between groups of fish become apparent. Look at the butterfly fish in Figure 4.21. What features do they all have in common? They are thin-bodied fish, many having a dark band across the eye. Each has a small mouth at the end of an extended snout and tiny bristle-like teeth. The three butterfly fish in Figure 4.21 all have these features and all belong to the genus *Chaetodon*. Look at Figure 4.21 and decide if you can see the similarities.

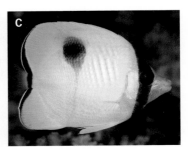

Figure 4.21 A The archer butterfly fish (*Chaetodon bennetti*), **B** the white spotted butterfly fish (*Chaetodon kleinii*) and **C** the limespot butterfly fish (*Chaetodon unimaculatus*). These butterfly fish share common characteristics and all belong to the same genus.

Q 4.17 Look at Figure 4.22. What distinctive feature do the bannerfish in the genus *Heniochus* seem to have in common?

The bannerfish shown in Figure 4.22 look rather like the butterfly fish in Figure 4.21, but they are sufficiently different to be grouped in a different genus, *Heniochus*. However, they do share a number of similarities with members of the genus *Chaetodon*, like the butterfly fish, so the two genera are grouped together in the same **family** – the Chaetodontidae (butterfly fishes).

Together with many other reef fish, members of this family all have thoracic, as opposed to abdominal, pelvic fins (that is, their lower fins are towards the front of the body). All families sharing this feature are grouped in the **order** Perciformes (perch-like), the largest order of fish, and all the orders of fish with bony skeletons are grouped together in the **class** Osteichthyes (bony fish). This class, along with all the other classes of animals with a dorsal spinal cord, is placed in the **phylum** Chordata – see Figure 4.23. All animals are grouped in the **kingdom Animalia**.

A taxonomic hierarchy is created. This is a series of nested groups or **taxa** (singular **taxon**), in which the members all share one or more common features or **homologies**. It is similar to the way in which you might arrange the folders and files in your work area of a computer (assuming you are that organised!).

The hierarchy of groups is:

- kingdom
- phylum (plural phyla)
- class
- order
- family
- genus (plural genera)
- species.

Figure 4.22 Bannerfish are related to the butterfly fish in Figure 4.21, but are sufficiently different to be classified in a different genus.

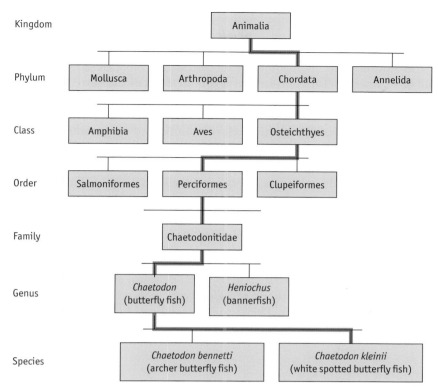

Kingdom — Animalia

Phylum — Mollusca, Arthropoda, Chordata, Annelida

Class — Amphibia, Aves, Osteichthyes

Order — Salmoniformes, Perciformes, Clupeiformes

Family — Chaetodonitidae

Genus — *Chaetodon* (butterfly fish), *Heniochus* (bannerfish)

Species — *Chaetodon bennetti* (archer butterfly fish), *Chaetodon kleinii* (white spotted butterfly fish)

Figure 4.23 Butterfly fish classification from kingdom to species (don't worry – you don't need to remember the names).

Butterfly fish are clearly animals. Along with many other organisms, we readily recognise them as members of the kingdom Animalia. In the same way, we can easily recognise the sea grass growing in the shallow waters around a coral reef as a member of the kingdom **Plantae**. Initially, taxonomists recognised only these two kingdoms: Plants and Animals. As scientists acknowledged that all organisms did not fit into these two kingdoms, further groups were formed. The separation of living things into five main kingdoms was proposed by R.H. Whittaker in 1959 and later modified by L. Margulis and K.V. Schwartz in 1982, giving the five kingdoms **Animalia**, **Plantae**, **Fungi**, **Protoctista** and **Prokaryotae**.

● kingdom Animalia – multicellular eukaryotes that are heterotrophs; organisms that obtain energy as 'ready-made' organic molecules by ingesting material from other organisms.

● kingdom Plantae – multicellular eukaryotes that are autotrophs; organisms that make their own organic molecules by photosynthesis (except a few parasites).

● kingdom Fungi – multicellular eukaryotes that are heterotrophs that absorb nutrients from decaying matter after external digestion.

● kingdom Protoctista – eukaryotes that photosynthesise or feed on organic matter from other sources but are not included in the other kingdoms; includes single-celled protozoa, such as *Amoeba* and *Paramecium*, and algae.

● kingdom Prokaryotae – prokaryotic organisms; includes the bacteria and blue-green bacteria (Cyanobacteria).

More recently, the system has been revised again; some groups from the kingdoms Fungi, Plantae and Protoctista have been placed together in a further kingdom, the Chromista. Their life cycles include motile cells with 'tinsel' like flagellae. This, and some other shared features, suggest that they are all more closely related to each other than to members of their original kingdoms. The kingdom Chromista includes such diverse organisms as the kelps (large brown seaweeds) and water moulds (including *Phytopthera infestans*, the species that caused the Irish potatoe famine). The remainder of the kingdom Protoctisa, now often called the kingdom Protozoa, is still under review.

As well as being a logical way to organise biodiversity, such as the shoals of brightly coloured fish around a coral reef, the groups have a deeper biological significance. Since members of a taxon share common features, it is likely that they also share a common evolutionary ancestor. This means that, in evolutionary terms, members of a taxon will be more closely related to each other than they will be to any organisms outside that taxon. In terms of the butterfly fish, members of the genus *Chaetodon* are all more closely related to each other than to members of the genus *Heniochus* and all of these fish are more closely related to one another than they are to members of other fish families. Diagrams of classification tell an evolutionary story, rather like a family tree.

New system of domains developed

In the 1960s, a scientist called Carl Woese, working at the University of Illinois, aimed to define the evolutionary relationships of prokaryotes. Woese pioneered RNA sequencing of bacteria, using the techniques available at the time. He used the sequences to construct phylogenies of groups of bacteria (Figure 4.24). A decade later, he noticed that one complete group of bacteria, the methanogens, completely lacked the sequences characteristic of bacteria. The RNA sequences showed up as different from all other bacteria, even with the crude early genetic sequencing techniques used by Woese. He supported his ideas with additional evidence, for example, the methanogens, unlike other bacteria, had no peptidoglycans (polymers of amino sugars cross-linked by peptide bonds) in their walls. They also had membrane lipids that differed from those of eukaryotes and other bacteria.

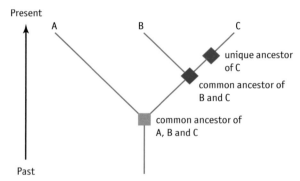

Figure 4.24 Phylogenetic trees show the evolutionary relationships between organisms based on molecular differences, the result of mutations accumulated over time.

A new group identified

In 1977, Woese proposed that this group belongs to a new category of organisms, the Archaea, forming a third branch of life alongside the prokaryotes and eukaryotes. Archaea may be related to some of the earliest life forms, existing before there was oxygen in the atmosphere. Today, they survive in extreme anaerobic environments, in places such as hot springs and salty lakes.

A paper published in the *Proceedings of the National Academy of Sciences* announced the new group. Individual scientists or research groups report their findings by publishing their work in scientific journals, in conference papers and on the Internet. This allows other scientists to see and comment on the data collected, the methods used and the interpretations. The scientific community carefully checks their data and interpretations of these data. This has to happen before new ideas are accepted as reliable scientific knowledge. This process of 'peer review' helps to detect invalid claims and adds weight to valid ones. But it can be difficult to get new theories accepted, particularly when they conflict with accepted theories.

Scientific community sceptical

Woese's theory was suggesting that most life is microbial and that scientists had been ignoring the existence of one of three major forms of life on Earth. Most of the scientific community were sceptical and ignored or dismissed Woese's ideas. Few said anything to Woese directly, or even responded in journals. Woese hated large meetings, so had few opportunities to argue in person on behalf of the Archaea. He seldom even attended the annual meetings of the American Society for Microbiology (ASM).

Evidence supports three domains

Luckily, the influential German microbiologist Otto Kandler accepted Woese's work. It was Kandler who organised the world's first Archaea conference in 1981, and influenced other scientists to accept the new group. He and other scientists published papers on the subject in respected international journals and eventually the theory gained acceptance as more and more scientists reported evidence showing the Archaea to be unique. In 1990, Woese, Otto Kandler and Mark Wheelis proposed a new system of classification, a universal phylogenetic tree based on three **domains**, the Archaea, the Bacteria and the Eukaryota (Figure 4.25). The organisms in each of the three domains contain RNA sequences that are unique to their domain.

> **CHECKPOINT**
>
> **4.3** Produce a series of bullet points to summarise how new data was used to produce the three domain classification, highlighting the role of critical evaluation by the scientific community.

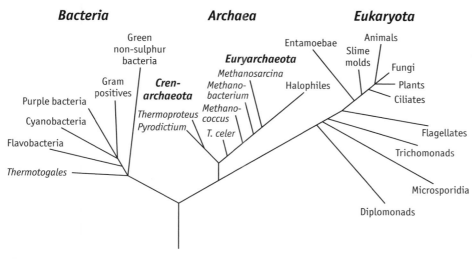

Figure 4.25 In 1990, Woese, Otto Kandler and Mark Wheelis proposed a new system of classification based on three domains, the Archaea, the Bacteria and the Eukaryota. It was derived from molecular differences caused by accumulation of mutations over time. By comparing the similarity of RNA between two species, it is possible to show their evolutionary relationship. Phylogenetic trees such as this suggest where species branched off from common ancestors.

The acceptance of the validity of Woese's classification was a slow process, but increasingly as new evidence emerged as a result of modern techniques, the scientific community critically evaluated the evidence, accepted his theory and recognised its full significance.

The final ratification of Woese's work came with completion of the genetic sequencing of *Methanococcus jannaschii* in 1996, which confirmed that the Archaea do indeed form a unique domain, with a closer evolutionary relationship to eukaryotes than to bacteria.

Classification is dynamic

Phylogenetic trees represent the evolutionary relationships based on the best existing evidence and they are constantly being revised as new data becomes available.

All modern classification systems attempt to show evolutionary relationships between species. DNA analysis often supports traditional Linnaean classifications, but not always. For example, in the traditional classification of the vertebrates, birds and reptiles are placed in separate classes. Anyone can see that birds differ from reptiles. They are warm-blooded and possess feathers, beaks and hard-shelled eggs. But DNA analysis places birds in the same taxon as turtles, lizards, snakes and crocodiles (Figure 4.26). This new classification is supported by fossil evidence, which shows that birds are descended from a type of dinosaur.

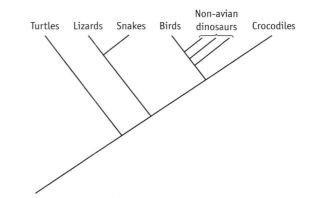

Figure 4.26 The group Reptilia, based on DNA analysis, includes animals from two separate Linnaean classes, reptiles and birds, showing that birds are as closely related to the different reptile groups as the reptiles are to each other.

> **ACTIVITY**
>
> In **Student Activity 4.11** you can investigate changing classification for yourself.

Convergent evolution causes problems when classifying

Phenotypic traits are observable characteristics, such as flower structure or bone structure. Organisms sometimes look similar only because they are adapted to similar conditions. For example, many plants living on windy mountain tops have hairy leaves (Figure 4.27), but they may belong to several different plant families. Dolphins and sharks look superficially similar, but they are not closely related (Figure 4.28). See also the mouse-like marsupial in Figure 4.3. These are examples of convergent evolution – unrelated organisms evolving a similar appearance as they adapt to the same conditions. In the case of the dolphin and the shark it is easy to see that they belong to different groups: a dolphin is a mammal and breathes air using lungs, whereas a shark is a fish and has gills. However, it is not always possible to be sure whether similarities are due to true relatedness rather than convergence. In addition, some organisms can have markedly different phenotypic traits from their close relatives. Using DNA analysis in classification overcomes these problems.

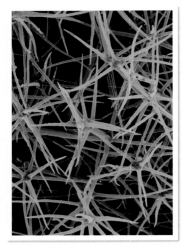

Figure 4.27 Scanning electron micrograph of leaf hairs of hairy mullen, typical of plants exposed to strong light and drying winds.

Figure 4.28 A dolphin and a shark share a similar shape, adapted to fast swimming and catching fish, but belong to different classes of vertebrates (mammals and fish).

Why classify?

Classification is a way of thinking about biodiversity. When used in partnership with genetic analysis, classification enables us to look at evolutionary relationships.

The ability to identify closely related species can be very valuable, for example, when trying to find new sources of chemicals with medicinal or other beneficial properties. An anti-viral drug was identified in the Australian tree species, the Moreton Bay Chestnut, but was found to be toxic. However, plants in a closely related genus were found to contain more of the drug in a less toxic form.

Figure 4.29 At first glance the zebras may all look the same, but on closer inspection there are differences between the individuals. Some of this variation will be genetic in origin.

Biodiversity within a species

Individuals within a species differ from one another – they show variation (Figure 4.29). You only need look around at the other members of your advanced biology group to realise just how much variation there can be. Topic 3 discussed how the appearance of an organism, its phenotype, is the result of interactions between its genotype and the environment. In all organisms that reproduce sexually, every individual (except for cases such as identical twins and cloned organisms) has a unique combination of alleles. This is **genetic diversity** and the greater the variety of genotypes the more genetically diverse the population. Genetic diversity allows the population to adapt to changing conditions and so should be conserved. If the population declines, some alleles may be lost and the genetic diversity decreases.

Sources of genetic variation

Where does all this genetic diversity come from? Topic 3 explained how meiosis results in genetic variation through independent assortment and crossing over. Random mutations also generate genetic variation (Figure 4.30). They do this by changing the base sequence of DNA in the cells of an organism, creating new alleles. Genetic diversity is increased with the addition of these new alleles to the gene pool. The gene pool consists of all the alleles of all the genes present in a population. In Topic 2 you saw how the deletion of three nucleotides was one of the mutations causing cystic fibrosis. You also saw how a single point mutation (alteration of one base) could result in the formation of malfunctioning haemoglobin, giving rise to sickle cell anaemia.

Most mutations have no effect on phenotype, some have harmful effects and some mutations can be beneficial to the organism. For example, mutations in houseflies that make them resistant to the pesticide DDT are an advantage when DDT is present. Even the sickle cell anaemia mutation can be advantageous. Heterozygotes who carry both sickle cell and normal alleles are more resistant to malaria. In countries where malaria is widespread, carrying the sickle cell allele gives a distinct advantage.

EXTENSION

Approximately one in seven human mutations is caused by jumping genes. To find out more read **Student Extension 4.3** – Jumping genes.

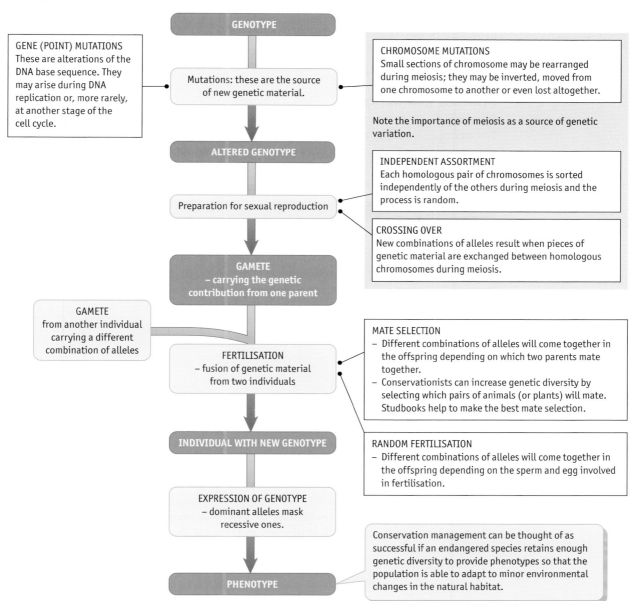

Figure 4.30 Sources of variation.

Genetic diversity may be visible

Sometimes, genetic differences within species show up clearly in the external phenotypes. For example, wild arum has a variable number of spots on its leaves and the spottiness increases in more northerly populations (Figure 4.31).

Much genetic variation has no visible effect on phenotype. It is made up of molecular differences that can only be detected using techniques such as gel electrophoresis of proteins: two alleles at a single locus produce slightly different protein products. However, these differences may be important in evolutionary terms.

Measuring biodiversity

Species richness

The simplest way to measure biodiversity is to count the number of species present in a given habitat. This is called **species richness**. A habitat is a place where organisms live, such as a woodland edge, a ditch, inside a rotting log in a rainforest, or a microhabitat such as the water-filled husk of a Brazil nut fruit. A clearing in a wood in southern England might have five butterfly species, but a similar clearing in a forest in South America may have ten times that many. By this measure we can say that the South American forest has more biodiversity than the English wood. In practice it may only be possible to list some of the species present, for example, vertebrates or vascular plants because many, such as algae or bacteria, are too small to see.

Species evenness

Species richness is a useful measure, but it takes no account of the population size of each species. For example, consider the hypothetical situation in Table 4.3. Wood A is clearly more diverse than woods B or C, because it has a greater number of species. But wood B is more diverse than wood C, though they both have five species. In wood B the levels of abundance of the individual species are more similar than in wood C.

Figure 4.31 Lots of natural variation reflects high genetic diversity within a species. Species with wide geographical ranges often show regional variations, sometimes with a gradual trend from north to south or east to west. Compare the differences between this wild arum and the one in Figure 4.17.

Species	Wood A	Wood B	Wood C
blue tit	25	20	50
robin	20	17	12
wren	11	23	3
blackbird	9	22	27
willow warbler	17	0	0
great tit	7	18	8
blackcap	4	0	0
song thrush	7	0	0

Table 4.3 Abundance of bird species (%) in three woods.

A community in which most of the species have similar abundances is said to have high evenness; no single species dominates the community. A highly diverse community would have high species richness and high species evenness. A community with an identical number of species, but which is dominated by one of them, is generally considered to be less diverse.

A very common species in a habitat is sometimes called the **dominant** organism. In English woods, oak trees are often dominant. This is partly because, over many years, humans have selectively encouraged oak trees because they provide good timber. In some habitats, a small number of animal species, such as pigeons and rats, can become dominant at the expense of the rest, and become pests. In species-rich ecosystems, such as rainforests, habitats usually show a more even species composition.

Q 4.18
Table 4.4 shows the tree composition in two imaginary British woods, one managed by humans and the other left in its natural state. Let us assume that each wood contained one hundred trees which were each identified. Compare the species richness and species evenness of the two woods and decide which has the greater biodiversity.

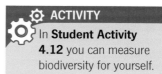

ACTIVITY
In **Student Activity 4.12** you can measure biodiversity for yourself.

	Managed wood	Ancient wood
Oak	50	15
Ash	10	15
Sycamore	25	0
Field maple	2	15
Hazel	3	20
Small-leaved lime	0	8
Wild service	0	9
Common hawthorn	5	0
Midland hawthorn	0	10
Holly	2	8
Beech	3	0
Total	100	100

Table 4.4 Tree composition of two imaginary woods.

A diversity index is a way of calculating a quantitative score for biodiversity that takes account of both richness and evenness. It can be used to compare the biodiversity of different habitats. One such index is calculated using the formula:

$$D = \frac{N(N-1)}{\sum n(n-1)}$$

where N is the total number of organisms of all species, n is the total number of organisms of each species and $\sum$ means sum of.

If, for example, you were investigating the diversity of a rock pool, you would identify all the species living in the pool and count the number of each species. Figures from one rock pool found 27 top shells, 305 barnacles, 4 limpets and 2 sea anemones. The diversity of the pool would be:

$$D = \frac{338 \times 337}{(27 \times 26) + (305 \times 304) + (4 \times 3) + (2 \times 1)}$$

$$= \frac{113\,906}{93\,436}$$

$$= 1.23$$

Q 4.19
Using the data in Table 4.4, calculate the diversity index for each of the two woods.

ACTIVITY
In **Student Activity 4.13** calculate diversity indices for contrasting communities.

Finding the hotspots

The world's biodiversity is not distributed evenly across the surface of the planet. In the last few decades, unexpectedly high levels of biodiversity have been discovered in unlikely places. For example, while coral reefs and rainforests are ranked as the richest ecosystems in the world, the biodiversity hotspot containing the largest proportion of the world's plant species is not, as you might expect, in the tropics. Instead, the plant hotspot is the Mediterranean Basin which contains one in ten of the Earth's plant species. More than half are **endemic**, that is found only in that area and nowhere else on Earth. Another example comes from ocean sea beds. These have long been regarded as very low diversity ecosystems. However, exploration of the muds of deep oceans has revealed a startling range of animals living in sediments previously thought to be lifeless.

The term 'biodiversity hotspot' was coined in 1988 by conservationist Norman Myers to describe areas of particularly high biodiversity. He identified 10 tropical forest regions containing between them 13% of all plant diversity in just 0.2% of the Earth's land area. Conservation International later adopted the idea as a way of focusing conservation effort on the most critical places. They have extended the list to include 25 regions collectively covering only about 2% of the Earth's land surface, yet containing more than 50% of all terrestrial species (Figure 4.32).

Q 4.20 Suggest why hotspots are identified only in terms of plant diversity.

Q 4.21 Describe a plausible explanation for the distribution of hotspots shown on the map in Figure 4.32.

WEBLINK

You can find out more about the hotspots by visiting the Conservation International website.

Figure 4.32 Biodiversity hotspots of the world. These particular hotspots are identified in terms of plant diversity. To be regarded as a hotspot according to these criteria, a region must have a minimum of 1250 different plant species. Crucially, it must also have 0.5% or more of total global plant diversity present as endemic species.

Measuring genetic diversity within a species

Genetic diversity can be measured directly or indirectly. The direct method is by DNA sequencing to determine the bases in a segment of DNA and thus to determine which alleles are present. Alternatively, the DNA can be cut into fragments and then separated using electrophoresis. Different alleles can be identified because they produce fragments of different lengths. Electrophoresis is studied in detail in Topic 6.

The genes from a group of individuals are sampled and the alleles present in this sample are recorded. The data are used to calculate an index that quantifies the genetic diversity. For example, the heterozygosity index is the proportion or percentage of genes that are present in the heterozygous form.

ACTIVITY

In **Student Activity 4.14** you can compare the heterozygosity indices for natterjack toad populations.

Conservation biologists may use genetic diversity indices to compare different populations when deciding which populations need to be conserved. If there is high genetic diversity within all the populations, then conservation effort can be focused on some populations without overall loss of genetic diversity.

4.4 Making use of biodiversity

Look around you and notice how many of the things that you rely on come from natural biodiversity, plants in particular (Figure 4.33). Adapted to survival in their own environments, the structures and products made by plants are often what we depend on as resources for our own survival.

WEBLINK
The diversity of plant uses can be seen at the Plants for a Future website.

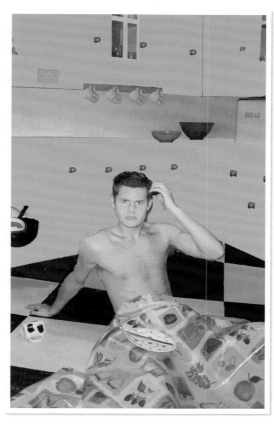

Figure 4.33 With and without plants.

How are plants adapted to survive?

Big and strong

Why do trees grow tall? Some plants lift their leaves on stalks or trunks so that they are above those of their competitors. If it can grow big enough, a tree can obtain massive amounts of photosynthetic energy and sink roots deep into the ground to take in the maximum amount of water and nutrients. The Brazil nut tree can reach a height of over 50 m and grows up to 20 m above the forest canopy to obtain light (Figure 4.34). If attacked by fungi, viruses or animals, part of their structure may be lost, but the rest can live on.

This is a risky strategy for survival since it can take a long time to grow to maturity. The Brazil nut tree does not produce fruit until it is about 45 years old. Yet it works; plants can build significantly bigger structures than any land animal. Redwoods can grow to 90 m, and 20 m palm trees can survive 100 mph winds. Such large structures must be strong enough to hold up their own weight and withstand the enormous wind forces on them. The same applies to upright annual plants, though they don't grow so tall because they only have one growing season in which to reach maturity.

Figure 4.34 Brazil nut trees can tower 20 m above the surrounding trees.

Building tall structures

All plants use three basic principles to build tall structures:

1 They produce strong cell walls out of cellulose, a polymer made from sugar molecules.

2 They build columns and tubes from specialised cells.

3 They stiffen some of these special cells with another polymer called **lignin**.

Trees add a ring of this stiffened (lignified) tissue each year and building up wood in this way allows trees to grow taller. The structures produced are not only strong, but also flexible, allowing trees to sway in high winds. We attempt to reproduce the properties of wood in composite materials, using more than one substance in combination. For example, concrete reinforced with steel is much stronger than either material alone.

DID YOU KNOW?

Reach for the sky

Like trees, tall buildings have to be strong enough so that they do not collapse under their own weight. Materials such as concrete provide strong walls that will bear this weight.

Tall buildings also have to deal with the horizontal force of wind. At their tops, most skyscrapers can move several feet in either direction, like a swaying tree, without damaging their structure. Any more movement would not only be uncomfortable for people inside, but would risk breaking up the concrete structures – these are brittle, unable to bend much without breaking.

Steel columns allow a degree of bending – they are much less brittle than concrete. However, a tall building made entirely from steel columns would eventually buckle and bend under its own weight. The solution is a steel framework combined with a concrete structure, providing both strength and stiffness (Figure 4.35). This allows the entire structure to move a little as one unit, like a tree.

Figure 4.35 Buildings have to withstand the same forces as tall trees. Architects copy plants when designing buildings, using concrete walls reinforced with steel, instead of cellulose walls reinforced with lignin.

In order to appreciate how plant cells are highly specialised for their function you first need to study a typical plant cell.

KEY BIOLOGICAL PRINCIPLE: HOW ARE PLANT CELLS DIFFERENT FROM ANIMAL CELLS?

Figure 4.36 shows the ultrastructure of a generalised plant cell. Decide how it is different from the animal cell shown in Figure 3.8 (page 108).

There are two fundamental differences:

● The plant cell has a rigid cell wall.

● The plant cell contains chloroplasts.

Chloroplasts are the site of photosynthesis, where energy from the Sun is used to make storage molecules. Starch is found in storage vacuoles in the cytoplasm called amyloplasts. In addition to the cell wall and chloroplasts in plant cells, there is often a large central vacuole surrounded by a vacuolar membrane (tonoplast).

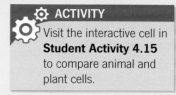

ACTIVITY

Visit the interactive cell in **Student Activity 4.15** to compare animal and plant cells.

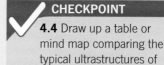

CHECKPOINT

4.4 Draw up a table or mind map comparing the typical ultrastructures of animal and plant cells.

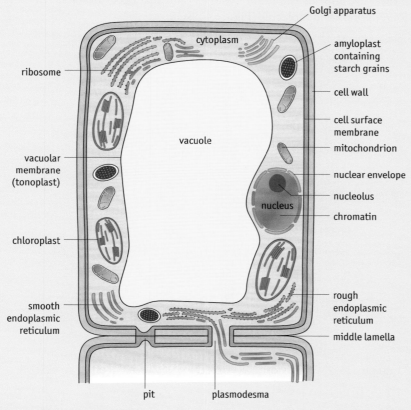

Figure 4.36 A generalised plant cell.

Parenchyma is a type of plant tissue found throughout the plant. The cells fill spaces between more specialised tissues and may themselves have certain specialised functions. For example, in roots they may have a role in storage and in leaves they contain chloroplasts and form the photosynthetic tissue (Figure 4.37). They also contribute to supporting the plant, see page 188.

 Q 4.22 The plant cell can be likened to a factory, with each of the organelles carrying out a special function. Look at the factory functions below and in each case decide which of the cell organelles performs an equivalent function:

(a) control centre where the instructions are kept to tell the factory floor what to make

(b) boiler room, harnessing energy to drive all the processes in the factory

(c) packaging hall, wrapping up and collecting products to be exported from the factory

(d) security screen, allowing in only those external raw materials that the factory needs

(e) warehouse storage facility within the factory.

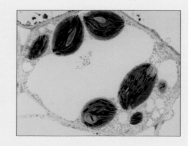

Figure 4.37 False colour electron micrograph showing parenchyma cells in the leaf contain many chloroplasts. Magnification ×2000.

Cell walls – the secret of their strength

Cellulose

A plant's strength comes in part from the thin **cellulose** walls of plant cells and the 'glue' that holds them together.

Cellulose is a polysaccharide.

It is a polymer of glucose, but the glucose it is made from is slightly different from that which forms starch. In Figure 4.38 the two forms of glucose are shown alongside each other. Can you spot the difference? You should notice that the –OH (hydroxyl) groups on the first carbon atoms are on opposite sides.

Cellulose is made up of β-glucose units. A condensation reaction between the –OH group on the first carbon of one glucose and the –OH on the fourth carbon of the adjacent glucose links the two glucose molecules (Figure 4.38B). A 1,4 glycosidic bond forms. In cellulose, all the glycosidic bonds are 1,4: there are none of the 1,6 glycosidic bonds that occur in starch. Because of this, cellulose is a long unbranched molecule (Figure 4.38C).

SUPPORT

You can remind yourself about condensation reactions and –OH groups by looking at the Biochemistry support on the website.

ACTIVITY

In **Student Activity 4.16** you can use the Biochemistry support on the website to see how β-glucoses join to form cellulose molecule, and also to compare the structures of starch and cellulose.

A Two forms of glucose

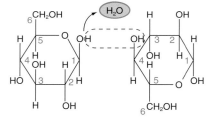

α-glucose   β-glucose

B Joining two β-glucose molecules

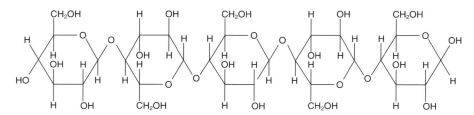

C Glucose chain – cellulose

Figure 4.38 A Cellulose is made up of β-glucoses joined together. Starch is composed of α-glucoses. **B** Formation of a 1,4 glycosidic bond between the two β-glucose molecules is only possible if one molecule is rotated through 180°. **C** A molecule of cellulose. Notice how the –OH groups project from both sides of the molecule due to the arrangement of the glucose molecules. Each alternate glucose is inverted to allow the 1,4 glycosidic bond to form.

Q 4.23 Look at Figure 4.38B.

(a) Describe how the two β-glucose molecules must be positioned relative to each other for a condensation reaction to take place between them.

(b) Explain why the bond between each pair of glucose molecules in cellulose is called a 1,4 glycosidic bond.

Q 4.24 Look back at Figures 1.34 and 1.35. Compare the structure of starch and cellulose, noting the similarities and differences.

Cellulose molecules form bundles: microfibrils.

Each cellulose chain typically contains between 1000 and 10 000 glucose units. Unlike an amylose molecule that winds into a spiral (Figure 1.39, page 34), the cellulose molecules remain as straight chains. Hydrogen bonds form between the −OH groups in neighbouring cellulose chains, forming bundles called **microfibrils** (Figure 4.39). Individually, the hydrogen bonds are relatively weak compared with the glycosidic bonds, but together the large number of hydrogen bonds in the microfibril produces a strong structure.

> **CHECKPOINT**
> **4.5** Look back at the structure of starch in Topic 1. Draw up a table comparing the structures and functions of starch and cellulose.

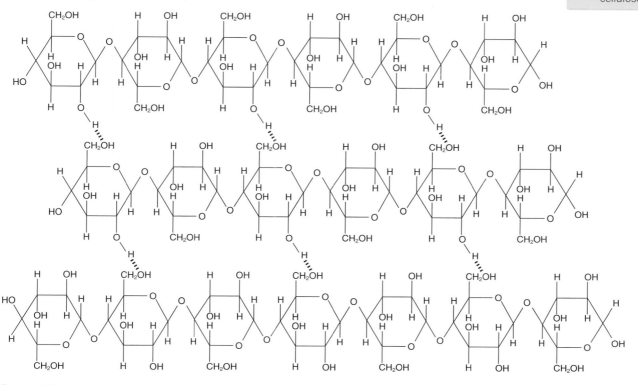

Figure 4.39 In cellulose, neighbouring chains of glucose molecules are linked by hydrogen bonds to form microfibrils.

If you look more closely at a plant cell wall (Figure 4.40), you can see that it is formed of microfibrils. These microfibrils are bundles of about 60–70 cellulose molecules. The microfibrils are wound in a helical arrangement around the cell and stuck together with a polysaccharide glue. Figure 4.40 shows how successive layers of the microfibrils are laid down at angles to one another, forming a composite structure.

The glue that holds together the microfibrils is composed of short, branched polysaccharides known as **hemicelluloses** and **pectins**. These short polysaccharides bind both to the surface of the cellulose and to each other, and hold the cellulose microfibrils together.

Pectins are also an important component of the middle lamella – the region found between the cell walls of adjacent cells. The pectins act as cement and hold the cells together.

The arrangement of the cellulose microfibrils within a matrix of hemicelluloses and pectins makes the cell wall very strong – rather like steel-reinforced rubber tyres. In this analogy, the hemicellulose matrix is the rubber, and the cellulose microfibrils are the reinforcing steel cables. This makes a strong, but pliable, structure. The microfibrils are laid down at different angles, which makes the wall strong and flexible.

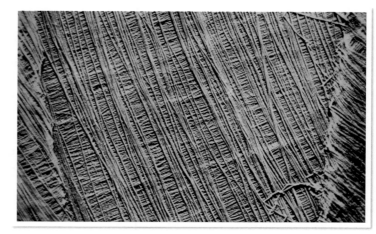

Figure 4.40 In the cell wall, layers of cellulose microfibrils are laid down at different angles.

Cross-linking

Looking at a pair of nylon tights under the microscope shows that there are tight chains running around the leg of the tights. These chains are joined more loosely vertically, so there is more 'give' lengthways. Net bags and elastic tubular bandages also show this uneven stretch. It is cross-links between the units that determine the amount of movement, and so the strength in any direction. Cross-linkages also strengthen scaffolding, electricity pylons and steel girders (Figure 4.41).

Q 4.24 Suggest how this idea can also be applied to cellulose in plants.

Figure 4.41 This structure relies on cross-linking to provide the structure with strength.

Pectins, jams and fruit juices

When fruits ripen they usually become soft (and eventually squashy) as the pectins holding the cells together are broken down by enzymes. The pectins become soluble, allowing the cells to shift when you squeeze the fruit.

In the fruit juice industry pectins are a nuisance. They thicken the juice, making it difficult to extract from the fruit, and slowing down filtration. Enzymes called pectinases are used commercially to break down the pectins, increasing yield and giving juices such as apple juice their clarity. Pectinases are also used to make baby food purée, and to peel the citrus fruits that come in tins.

Conversely, when making jam, pectins are a good thing – the more pectins there are, the better the jam sets, and the less fruit you have to use per jar. But if the fruit is too ripe the natural pectins will have broken down, and the jam will remain runny unless more pectins are added.

Crossing the cell wall

Cell walls do not separate plant cells completely. Narrow fluid-filled channels, called **plasmodesmata**, cross the cell walls, making the cytoplasm of one cell continuous with the cytoplasm of the next. Cell walls are also fully permeable to water and solutes. Can you identify the plasmodesmata in the cell wall shown in Figure 4.42? (Look at the photograph and decide before reading the caption.)

At some places the cell wall is thin because only the first layer of cellulose is deposited. The result is a **pit** in the cell wall. Plasmodesmata are often located in these pits, aiding the movement of substances between cells.

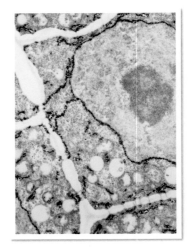

Figure 4.42 Photo of a section through a cell wall showing the channels between adjacent cells. The plasmodesmata are visible as parallel lines crossing the cell walls.

Tubes for transport and strength

To build a tall plant, such as a tree, some of the cells within the stem must be stiffened to provide mechanical support. At the same time, some cells must allow water and minerals (inorganic ions) to pass from the roots to the leaves; and others must distribute the products of photosynthesis around the plant.

There are three specialised types of cell of particular importance in fulfilling these functions. These are:

● xylem vessels – these form tubes for transport of water and minerals, and their stiffened cell walls help support the plant

● sclerenchyma fibres – columns of these cells with their stiffened cell walls also provide support

● phloem sieve tube cells – these form long tubes for transport of organic solutes, such as sugars and amino acids. They do not have a role in supporting the plant upright.

Where in the stem are these specialised cells?

There are three basic types of tissue found within plants. Figure 4.43 shows the general location of these three types: dermal tissue (epidermis), vascular tissue and ground tissue.

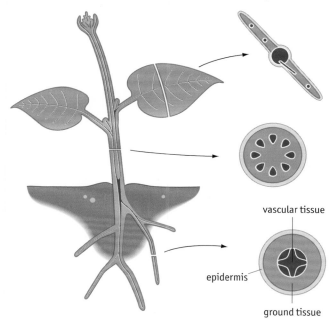

vascular tissue

epidermis

ground tissue

Figure 4.43 Where are the three basic types of tissue in the plant? The epidermis is a single layer of cells covering the entire outside of the plant. The vascular tissue, involved in transport, is surrounded by the ground tissue, which contains cells specialised for photosynthesis, storage and support.

Figure 4.45 Flowering plants, the angiosperms, can be divided into two classes depending on the number of seed leaves they have. Monocotyledon embryos have one seed leaf, their leaves have parallel veins and they rarely grow large because they cannot produce true wood. Dicotyledon embryos have two seed leaves and the veins on their leaves form a network pattern. They can produce wood and so include trees and shrubs. The grass is a monocotyledon. The clover is a dicotyledon. Both dicotyledons and monocotyledons have vascular bundles, but they are not arranged in the same way within the stem. In monocotyledons they are not arranged in circles, but are scattered through the stem.

Figure 4.44 overleaf shows the location of the **vascular tissue** within a cross-section of a stem from a dicotyledon (Figure 4.45). Notice how each **vascular bundle** contains **xylem vessels** and **phloem sieve tubes**. On the outside of the bundle are **sclerenchyma fibres**. In a young dicotyledon, the vascular tissue is in bundles towards the outside of the stem. In trees and shrubs these separate bundles merge to form a continuous ring as the plant ages. The xylem vessels carry water and inorganic ions up through the stem. The phloem transports products of photosynthesis, including sugars and amino acids, up and down the plant.

ACTIVITY
In **Student Activity 4.17** you can examine plant stems and locate the different tissue types yourself.

Q 4.26 Look at Figure 4.44C. Identify the locations of:

(a) a vascular bundle

(b) xylem vessels

(c) phloem sieve tubes

(d) sclerenchyma fibres.

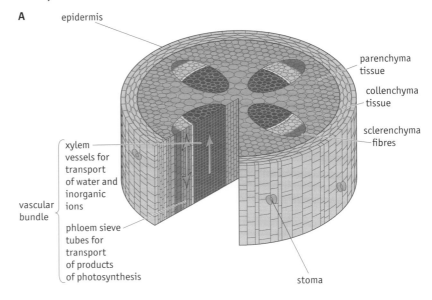

A

epidermis

parenchyma tissue

collenchyma tissue

sclerenchyma fibres

xylem vessels for transport of water and inorganic ions

vascular bundle

phloem sieve tubes for transport of products of photosynthesis

stoma

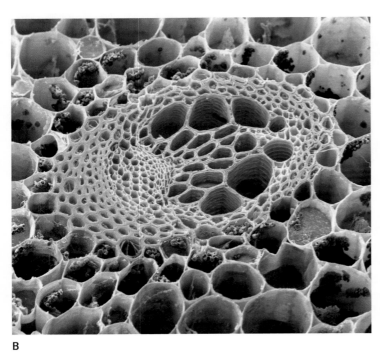

B

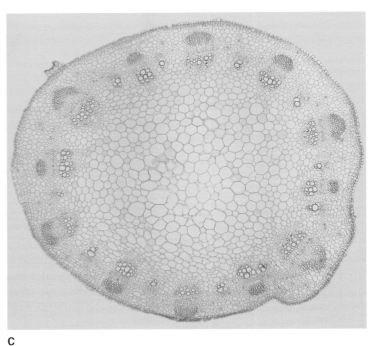

C

Figure 4.44 A The general arrangement of vascular tissue within the stem of a dicotyledon. The arrows represent the direction of flow through the xylem and phloem. **B** Coloured scanning electron micrograph of a vascular bundle of a buttercup, *Ranunculus repens*. The phloem sieve tubes are orange/yellow with the sclerenchyma fibres on their left and the xylem to their right. Magnification ×135. **C** A photomicrograph of a transverse section of a sunflower, *Helianthus*, stem.

Strong tubes

Tubes are useful things. They can carry materials and because they are hollow they are light. But tubes have a tendency to collapse in on themselves when put under tension. So they need to be reinforced if they are not already made of a strong material (Figure 4.46).

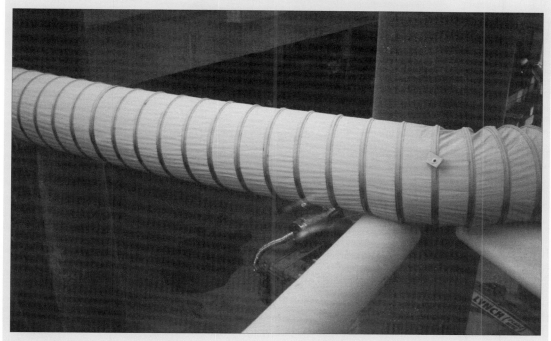

Figure 4.46 This large-scale vacuum tube uses spiral thickening to hold it open.

Xylem vessels and transport

Figure 4.44A shows a cross-section of a plant stem. The xylem vessels are made up of large cells with thick cell walls. They form a column of cells acting as tubes for the transport of water and mineral ions. In order to transport water, the cell walls have to be waterproofed. The plant produces another polymer – **lignin**. This polymer impregnates the cellulose cell wall and, as the cells become lignified, the entry of water and solutes into them is restricted. At about the same time, the tonoplast breaks down, and there is **autolysis** of the cell contents. During autolysis, the cell organelles, cytoplasm and cell surface membrane are broken down by the action of enzymes and are lost, leaving dead empty cells that form a tube.

The detailed structure of xylem vessels is shown in Figure 4.47. The end walls between the cells of the columns are lost or become highly perforated. Long tubes form as a result of this process and they are continuous from the roots of the plant to the leaves. The cellulose microfibrils and lignin in the cell walls of the xylem vessels give the tubes great strength and hence a role in supporting the plant.

Q 4.27 Explain how disruption of the tonoplast might result in the breakdown of the cell contents.

How is water transported through xylem vessels?

Xylem vessels are effectively fluid-filled tubes through which water moves upwards from the roots to the shoots. How is the water moved through these dead cells?

Water evaporates from all surfaces of the plant, mostly from the large surface area of the leaves. The majority of the evaporation occurs from the surfaces of the cells that line the substomatal

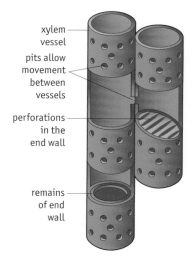

xylem vessel

pits allow movement between vessels

perforations in the end wall

remains of end wall

Figure 4.47 Xylem vessels cut open to show their detailed structure.

cavities in the leaves, as shown in Figure 4.48. Water diffuses out through the stomata down a **diffusion gradient**. Water evaporating from the plant in this way is known as **transpiration**. The water that leaves a plant leaf by transpiration is replaced by water absorbed through the roots. In the tallest trees water may move up to 100 m through the plant.

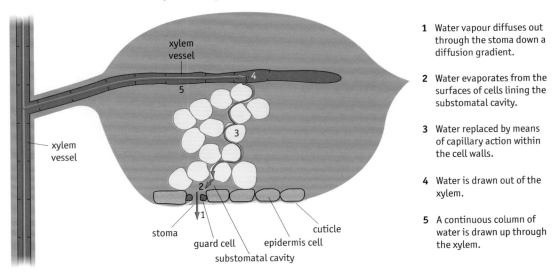

1 Water vapour diffuses out through the stoma down a diffusion gradient.

2 Water evaporates from the surfaces of cells lining the substomatal cavity.

3 Water replaced by means of capillary action within the cell walls.

4 Water is drawn out of the xylem.

5 A continuous column of water is drawn up through the xylem.

Figure 4.48 Water moves up through the xylem due to the evaporation of water from the surfaces of cells within the leaf.

ACTIVITY

The interactive tutorial in **Student Activity** **4.18** should help you understand the function of xylem in the transport of water through the stem. Use the Biochemistry support on the website to remind yourself about water and hydrogen bonding.

It is the evaporation of water from the cells in the substomatal cavities of the leaves that provides the force needed to draw water up a plant. The minute channels between the cellulose microfibrils in the cell walls act as tiny capillaries. Capillaries have the ability to draw water up by capillary action, which is caused by surface tension between the water and the capillary walls: see the Key Principles Box opposite. As water evaporates from the surfaces of the cell walls within the leaf, it is replaced by means of capillary action within the cell walls. This draws more water to the cell surfaces. The thousands of minute capillary-like channels in the cell walls inside the leaves produce a massive pull on the water behind them. This pull is thought to be sufficient to draw water from the xylem vessels and up the whole stem of the plant, even in large trees. Water moves up the xylem vessels and through the cell walls within the leaf in a continuous stream. This stream of water passing through the plant is known as the **transpiration stream**.

The energy for moving water through a plant comes from the Sun, which heats and evaporates water from the cells lining the substomatal cavity. Water is pulled up from the top of a plant rather like pulling on a piece of string. The water is under tension as it is pulled up. Xylem vessels would collapse inwards, like a strongly sucked straw, but for the reinforcing lignification of their walls. As in a straw, the water column does not break when it is pulled. This is because of the cohesive forces between water molecules, which are a result of hydrogen bonding – water in narrow tubes sticks together very strongly. Because the water is linked by cohesion and is pulled up under tension, this concept of water movement is known as the **cohesion-tension theory**.

What else do xylem vessels transport?

The movement of water through the xylem provides a **mass flow system** for the transport of inorganic ions. These are absorbed into the roots and are required throughout the plant.

Q 4.29 There is a very low concentration of inorganic ions in the soil solution surrounding roots. Using what you know about transport across cell surface membranes, suggest how the inorganic ions are transported into the root cells.

ACTIVITY

In **Student Activity** **4.19** you can investigate plant mineral deficiencies practically.

Nitrate ions are needed by plants in order to make amino acids. Amino acids contain one or more nitrogen atoms. Unlike us, plants make all their own amino acids from scratch using inorganic materials. By a sequence of enzyme-controlled reactions, the nitrogen from nitrate ions (transported in the xylem) is combined with organic molecules from photosynthesis to make all 20 amino acids. Plants cannot grow without nitrate ions: like all organisms, their cell cytoplasm is

KEY BIOLOGICAL PRINCIPLE: THE IMPORTANCE OF WATER TO PLANTS

Topic 1 described water as a polar molecule. This polarity results in the formation of hydrogen bonds between water molecules. Polarity is the cause of many of the properties of water, including its being liquid at room temperature (see page 8).

Cohesion and surface tension

Hydrogen bonding between water molecules results in strong cohesive forces between water molecules that keep the water together as a continuous column in xylem vessels (Figure 4.49).

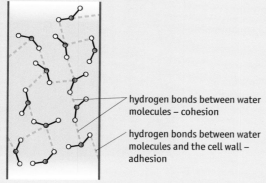

hydrogen bonds between water molecules – cohesion

hydrogen bonds between water molecules and the cell wall – adhesion

Figure 4.49 A sketch to show the forces of cohesion in a xylem vessel.

Surface tension at water surfaces is also partly caused by these cohesive forces between water molecules, which cause the surface layer of water to contract. This is useful for some small aquatic organisms, such as duckweed and pond skaters, which can be supported on this surface film. If a very fine tube is placed in water you can see the water move up the tube due to the adhesive and cohesive forces. This happens in the capillary-like tubes within cell walls in leaves.

Solvent properties

The solvent properties of water described in Topic 1 mean that dissolved substances can be transported around plants through the xylem and phloem. Once in cells, the dissolved chemicals move freely around in an aqueous environment and can react, often with water itself being involved in the reactions, for example, in hydrolysis and condensation reactions. The synthesis of sugars from water and carbon dioxide in photosynthesis is important to plants and to all organisms which depend on plants for their food supply. The details of photosynthesis and its role in the carbon cycle will be studied in Topic 5.

Thermal properties

Q 4.28 Recall from Topic 1 what specific heat capacity is and explain why it is very high for water.

You should remember that a large input of energy causes only a small increase in temperature, so water warms up and cools down slowly. This is extremely useful for organisms, helping them to avoid rapid changes in their internal temperatures and enabling many of them to maintain a fairly steady temperature even when the temperature of their surroundings varies considerably.

Density and freezing properties

Unlike most liquids, water expands as it freezes. As liquid water cools, the molecules slow down, enabling the maximum number of hydrogen bonds to form between the water molecules. These hydrogen bonds hold the water molecules further apart than in liquid water, making ice less dense than liquid water. Therefore, ice floats, enabling organisms to survive in liquid water under the ice in frozen oceans, ponds and lakes.

---

made largely of proteins, built by joining amino acids together (see Topic 2, page 63). Some other important biological molecules found in plants also contain nitrogen atoms, notably chlorophyll, nucleic acids, ATP and some plant growth substances.

If inorganic ions are not absorbed in sufficient amounts, the plant will show deficiency symptoms. For example, if the plant lacks magnesium it is unable to make chlorophyll and the older leaves become yellow between veins with reddish brown tints (Figure 4.50) – yellow leaves may also be a sign of nitrogen deficiency. A lack of calcium causes stunted growth due to the role of calcium ions in the structure of the cell wall and in the permeability of the cell membrane.

Phloem sieve tubes and transport

Figure 4.44A shows the location of **phloem sieve tubes** in the vascular bundle of a typical dicotyledenous flowering plant. Like a xylem vessel, a phloem sieve tube develops from a column of long, narrow cells in the growing stem. Unlike xylem vessels, phloem sieve tubes remain alive. However, the nucleus and most cell contents disintegrate during development, with only a few organelles remaining in a thin layer of cytoplasm close to the cell wall. This leaves most of the cell as a liquid-filled space, called the lumen. The end walls of each sieve tube cell contain holes that are aligned with those of the neighbouring cell to allow transfer of material between the fluid-filled lumen of adjacent cells. The perforated end walls are called **sieve plates** (Figure 4.51) and the section of a phloem sieve tube between sieve plates is called a **sieve tube element**. The lumen is continuous through the sieve plates at each end of the sieve tube element. Effectively, the lumen of a sieve tube is a vacuole that might stretch from a leaf at the top of a plant down to the end of a branch of its roots.

Figure 4.50 This potato plant is lacking in magnesium.

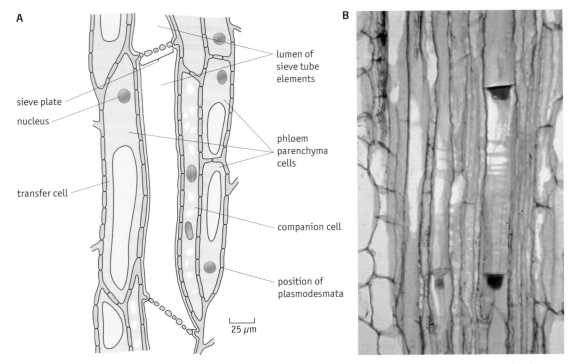

A

- sieve plate
- nucleus
- transfer cell
- lumen of sieve tube elements
- phloem parenchyma cells
- companion cell
- position of plasmodesmata

25 µm

B

Figure 4.51 A Phloem sieve tubes are associated with companion cells. **B** Light micrograph of phloem tissue, the sieve plates (red) separate the sieve tube cells. Companion cells are between the phloem sieve tubes.

The transport of organic molecules within the phloem is called **translocation**. The main substances transported are sugars (usually sucrose) and amino acids, produced in the leaves by photosynthesis. The fluid in the lumen moves along the sieve tube. In some sieve tubes the fluid flows from the leaves down to the roots where some sugars and amino acids are used for growth, and sugars may be used as a source of energy or converted to polysaccharide (such as starch) and stored. In other cases it may move from the leaves to the buds, developing flowers, and fruits and the seeds they contain, and this may mean moving upwards.

Q 4.30 Suggest where in a plant sugars would be transported to for storage.

Mineral ions taken up by roots travel to the leaves in the xylem vessels, but the nature of the way the xylem works means it is a one way process. When mineral ions move from one part of the plant to another they are transported in the phloem, for example, when various mineral ions move from old leaves to young growing leaves, or from daffodil leaves down into the underground bulb prior to going dormant. Growth regulators (the plant equivalent of hormones) and even viruses that cause plant diseases, may also travel in the phloem.

Alongside a sieve tube there is a long thin **companion cell** (Figure 4.51), which, unlike the sieve tube element, still has a nucleus, mitochondria, ribosomes and rough endoplasmic reticulum. These cells perform the metabolic functions that maintain the sieve tube.

Solutes are actively transported into and out of the phloem

Most plant cells are only a few cells away from a phloem sieve tube. The products of photosynthesis move out of the cells where they are produced, their source, and pass though these few cells until they arrive in a transfer cell next to a sieve tube (Figure 4.52). These specialised type of parenchyma cells help to actively transport the sugars, amino acids and other organic solutes produced as a result of photosynthesis into the sieve tube. When stored carbohydrates are mobilised they are also loaded into phloem in the same way.

The transfer cells are well adapted to this function: their cell walls and cell surface membranes have many small infoldings that increase their surface areas. They also contain numerous plasmodesmata linking their cytoplasm with that of adjacent cells. They also contain a lot of mitochondria to provide the energy necessary to load the solutes into the sieve tube.

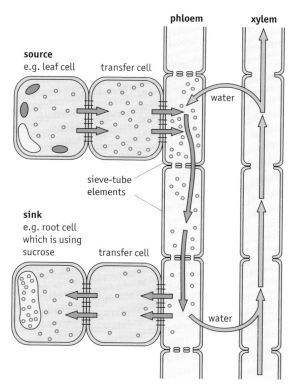

Figure 4.52 Simplified diagram showing how sugar molecules might be transported in phloem due to difference in pressure. At the source, sucrose (pink) is actively loaded into the phloem. As a result, water moves into the phloem by osmosis. At the sink, sucrose is actively unloaded from the phloem and water moves out of the phloem by osmosis. This creates a pressure difference which causes mass flow along the phloem.

Photosynthetic products are actively unloaded from the phloem sieve tubes in the parts of the plant where sugars are being used but not produced, for example, in the roots, buds, flowers and developing fruit. These are known as sinks.

A lot of plants survive the winter in a dormant state with starch stored in their roots, bulbs or tubers. When growth starts again in spring, the starch changes back to sugar. So, the root becomes the source and, for a short time, it is the buds and developing new leaves that become the sink, as the mobilised sugar moves through the phloem sieve tubes up the stem.

Q 4.31 Why is active transport necessary to load and unload the phloem sieve tube?

Mass transport in phloem sieve tubes

How are the sugar and other solutes moved along the phloem? Relying on diffusion would be too slow to meet the demands of the plant. Instead, a method of mass transport is thought to occur in the phloem. Loading of solutes to the phloem increases solute concentration, which draws water into the sieve tube by osmosis from adjacent xylem vessels. This increases the hydrostatic pressure inside the lumen of the sieve tube at the loading end. At sinks, solutes are unloaded, lowering solute concentration in the sieve tube. Water moves back into the xylem by osmosis lowering the hydrostatic pressure. The difference in pressure between the loading and unloading sites causes mass flow along the sieve tube: fluid carrying dissolved substances moves along the phloem sieve tube from high to low pressure.

Q 4.32 Explain how the mass transport system in plants is **a** similar and **b** different to the circulatory system found in mammals.

This theory of mass flow in phloem was first suggested in 1927 and although experimental evidence supports the model, there are concerns expressed about how this model could generate a pressure difference sufficiently large to allow transport in tall trees. Although some studies suggest that a large pressure difference is not necessary, the mechanism of phloem transport, particularly in trees, is not yet fully understood.

⚙ ACTIVITY

In **Student Activity 4.20** you can look at some of the evidence in support of the role of phloem sieve tubes in translocation.

Xylem and sclerenchyma for support

In addition to a role in plant transport, xylem vessels are important in supporting plants. Lignin not only waterproofs the cell walls, it also makes them much stiffer and gives the plant much greater tensile strength. Instead of forming a uniform layer on the inside of the xylem cell walls, lignin is often laid down in spirals or in rings as shown in Figure 4.53.

Phloem sieve tubes, companion cells and transfer cells do not have lignified cell walls and do not play any part in providing support. However, xylem vessels are not the only cells that become impregnated with lignin. The sclerenchyma fibres associated with vascular bundles in the stem and leaves also have lignin deposited in their cell walls. 'Sclerenchyma' comes from the Greek *scleros* meaning 'hard'. As with xylem vessels, the sclerenchyma fibres die once lignified, leaving hollow fibres (Figure 4.54). The strength of these fibres varies in different plant species depending on the length of the fibres and the degree of lignification.

The taller a plant grows, the greater the proportion of its stem that becomes lignified. In a tree, this is the majority of the trunk: the living parts are towards the surface (under the bark) and grow new layers each year. These layers are the annual rings. In annual plants the lignification is confined to the vascular bundles. The plant stem relies on tightly packed, fully turgid parenchyma cells to maintain its shape and keep it erect. A **turgid** cell is one that is completely full, with its cell contents pressing out on the cell wall. Turgor supports the leaves of all plants including those of trees. If a cell loses water, turgor is lost. If a high proportion of a plant's cells lose their turgidity, the plant wilts.

Useful plant fibres

Humans use plants in many ways. People have been using plant fibres for thousands of years in order to make products such as clothing, rope, floor coverings, paper and many more. Plant fibres can be used in these ways because they are:

- long and thin
- flexible
- strong.

How do we extract fibres from plants?

To obtain fibres we must take the plant apart. This can be done mechanically by pulling out the fibres or by digesting the surrounding tissue. Fortunately for us, cellulose – and particularly cellulose combined with lignin – is very resistant to chemical and enzymic degradation, whilst the polysaccharides that hold the fibres together can be dissolved away.

The more lignin there is present, the harder it is to separate fibres. So to produce fibre pulp from trees, caustic alkali is required. For flax (Figure 4.55) and other suitable plants, a milder treatment is used and in some traditional processes the stems are piled in heaps, allowing bacteria and fungi to do the work. This process, and its more modern chemical and enzymic equivalents, is called 'retting'.

Q 4.33 What do you think makes plant fibres a success commercially?

Not just textiles …

Fibres have many other uses in addition to the textile industry. For example, mats of fibres are used to absorb heavy metals and also hydrocarbons from polluted water (Figure 4.56).

Plant fibres can also be added to other materials to form biocomposites. For example, when oilseed rape (also known as rapeseed) fibres are mixed with plastic, the resulting material is stronger than plastic alone. These biocomposites are also renewable, more biodegradable and can be easier and safer to handle than composites containing artificial fibres.

Researchers at Warwick University have built a car from biomass. Their Eco One sports car has tyres made from potatoes, brake pads made of ground cashew shells and a body built from hemp and rapeseed. Only the car's steering-wheel, seat, electrics and chassis are made from conventional

Figure 4.53 The lignin thickening in xylem vessels can take different forms as shown in this photomicrograph.

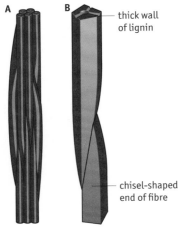

thick wall of lignin

chisel-shaped end of fibre

Figure 4.54 A Several sclerenchyma fibres. **B** Part of two fibres shown at a higher magnification.

Figure 4.55 Scanning electron micrograph of fibres from a flax plant (*Linum usitatissimum*) used in linen fabric.

Figure 4.56 Oil pollution control. Natural fibres absorb hydrocarbon pollution from water.

materials. The car runs on a special biofuel made entirely from sugar beet and fermented wheat. It is not the first time such a car has been produced: Henry Ford produced a car made entirely from hemp in 1941.

Chemical defences against attack

A plant cannot move much, it cannot run and it cannot hide, so it is an easy target for animals, bacteria and fungi. The agouti eats most of the Brazil nut tree's nuts, but the tree benefits as it helps disperse its seeds. However, many plants have adaptations that provide chemical defences to repel and even kill animals that feed on them. One strategy is to produce a chemical that is distasteful or even toxic. If the animal takes a bite and the taste is offensive (usually bitter, Figure 4.57), the animal is deterred from feeding further. If its chemicals kill the predator, the plant will avoid future attacks.

DID YOU KNOW?

Pyrethrum – natural insecticides

Pyrethrum is a kind of chrysanthemum and a member of the daisy family (Figure 4.58). Pyrethrum has the remarkable property of having no known pests or diseases. It contains chemicals that are extremely toxic to insects. Pyrethrum is grown commercially in Kenya, where flowers are picked, dried in the sun, then bagged and transported to factories. Here they are crushed into powder and further refined to extract the active ingredients: various pyrethrins.

Pyrethrins are used in many insecticide recipes in their native home, Kenya. For example, they are used in sprays for fruit and flowers, flea treatments for cats and dogs, and the reduction of tick infestation in cattle.

Figure 4.57 The bitter taste in tea is caused by the presence of tannins in the leaves. These also have the effect of 'gumming up' the mouthparts of aphids and so protect the tea plant from this form of insect attack.

Figure 4.58 *Tanacetum coccinium*, the pyrethrum plant.

Fortunately, these compounds are much less toxic to vertebrates than to insects, but they are not entirely harmless and must be handled with care. Pyrethrins have the advantage that they are unstable after spraying and rapidly decompose to harmless residues.

One alternative to using pyrethrins is to use the plants themselves. Organic gardeners use chrysanthemums and marigolds as 'companion' plants – planted amongst vegetables, they act as a natural insect repellent. You might try growing some chrysanthemums and testing them on flies.

KEY BIOLOGICAL PRINCIPLE: BACTERIAL GROWTH

Bacteria reproduce asexually in a process called binary fission; the circular DNA replicates and new cell content is synthesised before new cell wall forms to divide the cell into two roughly equal halves. You have probably seen video clips of one bacterium dividing into two, each splits giving rise to four, then eight, sixteen, thirty two and so on. This is exponential growth, in 24 hours a species like *Escherishai coli* with a generation time of 20 minutes would produce 2^{72} bacteria from the one original bacterium.

With bacteria reproducing so quickly why are we not completely overrun? The reason is that these rates of growth only occurs under ideal conditions, when there is sufficient nutrients, optimum temperature, optimum pH, no build up of toxic waste products, and sufficient oxygen if the bacteria relies on aerobic respiration. In reality conditions are rarely idea. If bacteria are grown in the laboratory in a closed system, like a conical flask, a typical bacterial growth curve is obtained (Figure 4.59A). Notice that log values of the number of bacteria are used as this produces a straight line rather than an exponential shaped curve (Figure 4.59B). The growth curve shows four phases: the lag phase - cells are adjusting to the conditions (e.g. enzyme synthesis); the exponential phase (also known as the log phase) - cells are dividing

exponentially at fastest rate possible for conditions; the stationary phase – growth is limited by lack of food, build of toxins, changing pH (balance between cell division and cell death); death phase – number of cell deaths is greater than cells formed.

Q 4.34 **(a)** Estimate how many bacterial cells will be produced in 24 hours by division of a single cell of **(i)** *Staphlococcus aureus*, which has a generation time of 30 minutes, and **(ii)** *Mycobacterium tuberculosis*, which has a generation time of about 15 hours?

(b) If a culture contains a million cells how many cells will it contain after four generations of every cell? Give your answer in standard form.

Q 4.35 Work out **(a)** how many cells will be produced from a single bacterium after **(i)** 12 generations and **(ii)** 14 generations (you can use the x^y button on your calculator when finding this value) **(b)** what $\log_{10}$ value would be plotted on Figure 4.59B for each of these cell numbers.

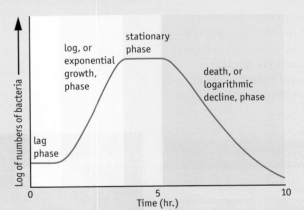

Figure 4.59 A Typical bacterial growth curve. **B** Change in bacterial populations follows a logarithmic progression, two cells, four cells, eight cells, etc.; after five generations there will be $2^5 = 32$ cells, after 10 generations, $2^{10} = 1024$. The exponential curve on the right shows this change in bacterial numbers; notice how this is plotted on a log scale to fit it onto the graph. Using log values the numbers can be plotted on a standard scale to produce a straight line.

Natural antibacterials

Plants sometimes store toxic compounds in hairs on the surfaces of their leaves. This is very obvious in the stinging nettle, which most of us avoid, but not so apparent in mint, which produces the chemicals menthol and carvone. These chemicals are toxic to microbes and some insects, but attractive to us as flavouring in foods or tea. Even humans experience the numbing effect of mint (whilst brushing your teeth, for example), but to microorganisms this can be lethal.

Garlic extracts have been found to destroy bacteria, such as *Campylobacter* and *Helicobacter*, which cause intestinal infections. This is potentially important as some strains of the bacteria are now resistant to widely used antibiotics, such as penicillin. The active ingredient in garlic is allicin; this is known to interfere with lipid synthesis and RNA production. Allicin is only produced when the plant is cut or damaged. Its inactive precursor, alliin, is converted into the active form by the enzyme alliinase.

Some studies have shown that certain parts of a plant tend to have greater antibacterial properties than the rest. These are typically the seed coat, fruit coat, bulb and roots.

⚙ **ACTIVITY**
You can investigate the antibacterial properties of plants in **Student Activity 4.22**.

Q 4.36 Suggest why the parts of the plant listed above need to produce the most antibacterial chemicals.

Medicines from plants

Not surprisingly, many plants contain poisons or produce them rapidly as a response to wounding. However, 'poison' is a relative term and relates to the dose necessary to cause harm to an organism. Clearly, if a chemical can kill pathogenic (disease-causing) microbes or malignant cancer cells at a dose level that leaves humans alive, then this 'poison' is a likely medicine. An enormous number of medicines are derived from chemicals originally discovered in plants.

The World Health Organization estimates that 75–80% of the world's population uses extracts from plants as medicines. Many of the common medicines we use were originally plant-derived, such as aspirin (salicylic acid derived from an extract of willow bark), and morphine and codeine (both derived from opium poppies).

Digitalis and drug development

Foxgloves and dropsy

Foxglove leaves are poisonous when eaten by humans and other animals. They have a strong, bitter taste that serves as a warning. The symptoms of poisoning are dizziness, vomiting, hallucinations and heart failure, caused by an irregular heartbeat. However, it is the effect on the heartbeat that made it a traditional folk remedy when used in moderation.

Figure 4.60 The foxglove, *Digitalis purpurea*, has been used in medicines for hundreds of years.

The foxglove (Figure 4.60) was known for centuries to have medicinal qualities and in particular was used to treat a condition known as dropsy. Dropsy, now called oedema, happens when fluid accumulates in the body tissues (see Topic 1, page 27). This process is painful and can cause a slow death. Oedema is usually caused by heart or kidney problems. A fast and irregular heartbeat is one of the signs.

Because the blood pressure is raised, tissue fluid fails to return to the capillaries and accumulates in the patient's feet, legs and organs, causing them to swell up. Eventually, the patient may drown as fluid fills up their lungs. So it was not surprising that a herb known to relieve this serious condition was used as a remedy. However, it was not until a country doctor called William Withering (Figure 4.61) published *A Treatise on the Foxglove* in 1775 that it became an accepted form of medicine.

Withering had heard of the foxglove's curative properties for dropsy, but his attention was focused when he met Mrs Hutton, a 'wise woman' who was showing signs of the disease. She assured him that she would be all right after she'd had a cup of her special tea. Imagine Withering's surprise when he visited her again and found that she had recovered.

Figure 4.61 William Withering.

Talking to Mrs Hutton, Withering found out that foxglove was amongst the 20 or so herbs used in the potion. He suspected that it was something in the foxglove that was the active ingredient. Later, Mrs Hutton sold her recipe to Withering who began to investigate the plant further. One of his first patients was a brewer suffering from swollen limbs and an irregular heartbeat. After a few doses of Withering's 'digitalis soup' he became healthy and his pulse became 'more full and regular'. Unfortunately, his next patient, an old woman, nearly died from the treatment, so Withering gave up his investigations.

Getting the dose right

After moving to Birmingham General Hospital, which received many patients suffering from dropsy, Withering was persuaded to renew his investigations. Studying 163 patients, he discovered and recorded side effects of digitalis. These included nausea, vomiting, diarrhoea and a rather strange green/yellow vision. A sign that the patient was recovering was the production of a large quantity of urine.

Withering realised that getting the dose right for the patient was of vital importance. He applied a standard procedure to discover the correct dosage for each patient. He slowly increased the dose until the patient started to have diarrhoea and vomiting, and then reduced it slightly: this would be the most effective dose. Withering meticulously recorded all his results and after ten years wrote his book about the medicinal properties of the foxglove. This helped change the face of medical practice forever. We now know that the active ingredient in the foxglove is the chemical called digitalin.

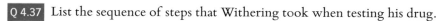
Q 4.37 List the sequence of steps that Withering took when testing his drug.

ACTIVITY
You make a comparison between William Withering's approach to drug development and that of the drug companies in **Student Activity 4.23**.

Drug testing today

Today, a potential new drug must pass a series of tests if it is to be developed into a new product. It has to be proven to be effective, safe and capable of making a profit. It can typically take 10–12 years and cost over US$1 billion to develop a drug.

Potential substances are analysed and the active ingredient (the drug compound that may bring about a cure) is identified and copied so that it can be manufactured synthetically. Slight variations of the chemical structure are made just in case they might have a better effect. (Nowadays many new drugs are 'discovered' through computer modelling of their chemical structures.)

A series of trials of the compound now begins. There are five stages in all. The first is pre-clinical testing, followed by three phases of **clinical trials** and finally after licensing trials. Table 4.5 describes these five stages.

Pre-clinical testing	Animal studies and laboratory studies on isolated cells and tissue cultures assess safety and determine whether or not the compound is effective against the target disease. These tests can take several years to complete. Thousands of chemicals go through pre-clinical testing, but only a handful are ever approved for clinical trials on humans. Animal trials form the basis of applications for clinical trials on humans, which are authorised by the Medicines and Healthcare products Regulatory Agency (MHRA), an independent body of scientists, doctors and members of the public.
Clinical trials – phase I	A small group of volunteers are told about the drug and given different doses. These volunteers are normally healthy, but there are circumstances where drugs are first tested on sick patients. The trial confirms whether or not the compound is being absorbed, distributed, metabolised and excreted by the body in the way predicted by the laboratory tests. The effects of different doses are monitored. In the UK, a review of the data collected is made by the UK MHRA.
Clinical trials – phase II	Small groups of volunteer patients (100–300 people with the disease) are treated to look at the drug's effectiveness. If the results are promising, phase III trials are set up.
Clinical trials – phase III	A large group of patients (1000–3000 people) is selected and divided randomly into two groups. One group is given the compound being investigated. The second is given an inactive 'dummy' compound known as a **placebo**. If there is an existing treatment for the disease, the standard treatment is given rather than a placebo. It is important that neither the patients nor the doctors know who is having the compound under investigation and who is having the placebo or standard treatment. This is known as a **double-blind randomised controlled trial**, and is considered the 'gold standard' of valid testing.
	If the compound being investigated is effective, then the results will show a statistically significant improvement in the patients receiving the treatment compared with patients given the placebo or standard treatment. The tests also look for any adverse reactions in the patients.
	The way is now open to license the compound as a drug, after which it can be marketed.
After licensing	Trials continue to collect data on the effectiveness and safety of a new drug after the drug has been licensed.

Table 4.5 Stages in drug testing.

Q 4.38 (a) Explain why the review team in the phase I trials needs to be independent.

(b) If phase II is a success, why is phase III needed?

(c) Is it ethical to give some patients a placebo, when it is known that the compound is likely to have a beneficial effect?

(d) Sometimes patients on a placebo will show an improvement in their condition. Suggest why this may be so.

Seeds for survival

Having successfully survived the onslaught of all sorts of herbivores, the plant's next challenge is to make sure enough of the next generation also survive. Flowering plants have achieved this by packaging a miniature plant in a protective coat with its own food supply; we call them seeds. Inside the seed the embryo remains dormant until conditions are suitable for restarting growth (Figure 4.62). The brazil nut is the brazil nut tree's seed.

Seeds are vital to the survival of a plant. They are adapted to ensure they:

● protect the embryo

● aid dispersal

● provide nutrition for the new plant.

What's in a seed?

In flowering plants the ovule is fertilised by the nucleus from a pollen grain and develops into the seed (see Topic 3 page 118).

The outer layers of the ovule become lignified forming a tough seed coat that protects the embryo within the seed. The surrounding ovary develops into the fruit (Figure 3.18B, which often has an important role in seed dispersal.

In some species the stored food in the seed remains outside the embryo in storage tissue called endosperm, as shown in Figure 4.63. This is common in monocotyledons, for example, cereals. Seeds of this type are called endospermic. In many dicotyledons the embryo absorbs the stored nutrients from the endosperm and the food is stored in the seed leaves (cotyledons), which swell to fill the seed (Figure 4.64). In some seeds, including Brazil nuts, there are no apparent cotyledons and the food is stored in the hypocotyls, the developing stalk.

Seeds come in all sizes and shapes, most of which are appropriate for wide dispersal. This helps offspring to avoid competing with their parent plant or with each other and it also lets plants colonise new habitats. Most species are adapted to take advantage of just one method of dispersal – wind, animal, water or self-dispersal.

When conditions are suitable and any dormancy has been broken, the seed takes in water through a small pore in the seed coat. Absorbing water triggers metabolic changes in the seed.

Figure 4.62 The seed ensures that the embryo it contains has the best chance of germinating and establishing a new plant.

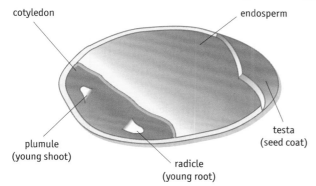

Figure 4.63 The internal structure of an endospermic seed.

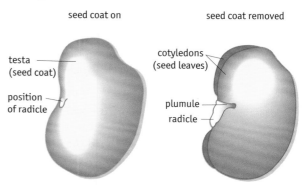

Figure 4.64 Opening up a broad bean reveals the cotyledons (seed leaves), plumule (young shoot) and radicle (young root) that make up the embryo.

Enzymes are produced that mobilise the stored food reserves. Maltase and amylase break the starch down into glucose, which is converted to sucrose for transport to the radicle and plumule. Proteases break down the proteins in the food store into amino acids; lipases break down the stored lipids into glycerol and fatty acids.

What can we do with starch from seeds?

Although cereals are grown mostly for human food and animal feed, their carbohydrate polymers and oils also have major industrial uses.

Starch is easy to extract from plants, particularly from seeds such as wheat and other cereal crops. This is because starch is in granules, which do not dissolve in water, but can be washed out. In wheat, the protein remaining is gluten, which is a rubbery mass. The elastic properties of gluten are vital in bread-making, but are not much used in industry.

We all eat starch, but it has many other uses. Starch is found in a wide range of products including adhesives, paints, textiles, plaster, insulating material and toiletries, such as conditioners, mousses, sun screens and anti-perspirants.

When starch granules are heated in water they suddenly swell, absorb water and thicken the liquid. This is 'gelatinisation' and this thickening process is the basis of both custard and wallpaper paste. An enormous amount of starch is used in paper coatings and cloth treatment. A starch mixture applied to the surface is gelatinised and then cooled, allowing bonds to form between the starch molecules stiffening the fabric. The addition of water reverses the stiffening.

Super-absorbents

If starch is chemically cross-linked before it is gelatinised then particles are formed that can be dried. When rehydrated, these particles can take up large amounts of water. Such cross-linked starch can now be found as super-absorbents in some nappies.

Starch foam

The temperature at which starch gelatinises depends on the amount of water present. At water contents of less than 10% the gelatinisation occurs at much higher temperatures: above the boiling point of water. If the pressure is raised at the same time, the starch forms a plastic mass; if the pressure is suddenly released, for example, when the seed coat ruptures during cooking, then steam forms and the starch 'puffs' into an expanded structure.

Puffed wheat breakfast cereals, corn snacks and starch-based foam packaging (Figure 4.65) are all made using this physical process, though corn snacks and starch-based foam are made in an extruder where a machine applies the pressure, rather than the seed coat. As hot starch leaves through a small exit hole the pressure that has built up inside is released, causing the starch to expand into foam as super-heated water turns to steam.

Starch-based packaging can now be used instead of polystyrene, polyethylene or other oil-based plastics. Plant-based plastics have also been developed that are made by fermentation of sugars from wheat, sugar beet, potatoes or agricultural waste. This means that supermarket food packaging, such as a foam tray covered in plastic film, can now be made from starch.

What can we do with vegetable oils?

Seeds are a rich source of oils, which we regularly use in cooking. Oils are not only used for food though – many other industrial uses have been developed.

Fuels

There is nothing new about using vegetable oils instead of petroleum-based products for motor vehicles. Castrol, the engine oil, was originally derived from the castor bean. Dr Rudolf Diesel developed the first diesel engine to run on vegetable oil and Diesel's engine was demonstrated at the 1900 World Exhibition in Paris using peanut oil. Biodiesel produced today can be used in unmodified diesel engines alternating with petroleum diesel. It produces less sulphur dioxide than diesel and less carbon dioxide when you take into account the carbon dioxide used by the plants grown to produce it.

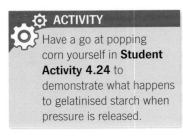

ACTIVITY

Have a go at popping corn yourself in **Student Activity 4.24** to demonstrate what happens to gelatinised starch when pressure is released.

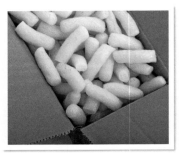

Figure 4.65 Foam packaging is commonly used to protect goods such as computers and telephones from knocks during transit and, to some extent, from large temperature fluctuations.

Across the UK there is a growing market for biodiesel. It is possible to buy biodiesel made from waste vegetable cooking oil or from oil crops, such as rapeseed. It is available either as 100% biodiesel, or as a blend with fossil fuel diesel, for use in standard diesel engines.

Sustainability

The use of oil-based plastics and fuels is not sustainable for several reasons.

- Burning fossil fuels contributes to a net increase in atmospheric carbon dioxide concentration, which is likely to contribute to climate change.

- Oil reserves will eventually run out.

- Plastics generate non-biodegradable waste, creating major waste disposal problems.

The use of plant-based products should help reduce these problems. Although burning fuel made from vegetable oil also produces carbon dioxide, this carbon dioxide has been removed from the atmosphere relatively recently when the crop that produced the oil was grown. The carbon dioxide released will be about the same quantity as the amount fixed, so there is no net change in carbon dioxide in the atmosphere, This is not the case when burning fossil fuels, which release carbon stores built up over long periods of time. The EU target is for 10% of our vehicle fuel consumption to be from renewable energy sources by 2020, with a significant contribution from biofuels. However, the EU is limiting the amount that food crop-based biofuels can count towards this target; instead encouraging the use of agricutural waste and non-food crops. This is to prevent the change of land use from food production to biofuel crops. The demand for food remains so it results in expansion of land used for food crops with an associated release of carbon dioxide with the change in use.

Sustainability is not guaranteed by simply switching to products made from plants that were alive very recently rather than fossil plants. We need to consider the source of the plant product, and the energy used and pollution created during the production and transport of the product.

Using paper bags instead of plastic ones in supermarkets might seem like a good idea. But transport of the heavier bags to the stores uses more fuel and, if the bag gets wet, the shopping ends up on the ground as the bag disintegrates. A plant-based plastic bag that remains intact when wet, is light and will also biodegrade is a more sustainable alternative.

Does biodegradable packaging solve our waste problem? Most bacteria involved in degradation require oxygen, so what happens if the packaging is buried deep in a landfill site? The answer is remarkably little – newspapers have been retrieved from deep landfill after ten years and have still been readable. Even a frankfurter had not degraded. Closer to the surface, biodegradable packaging will degrade, though sometimes methane is generated, which contributes to global warming.

DID YOU KNOW?

Brazil nuts and sustainability

The Brazil nut is the only nut that is normally harvested from the wild. Attempts have been made to grow Brazil nut trees in plantations, but their productivity is very low. This is probably due to the lack of orchid bee pollinators. Production is highest in undisturbed rainforest. Brazil nut production is important to the local economy, but there are concerns that in many areas the harvesting is not sustainable. A study found that in a sample of 23 populations that had been heavily harvested, young trees were missing from the populations. Over-exploitation of the nuts threatens the future of the trees and all the organisms that depend on the trees. Harvesting frequently disturbs the forest and affects productivity.

Although the long-term monetary value of the forest in terms of production of non-timber products is estimated to be greater than the value of the timber itself, there is widespread deforestation. The Brazil nut tree is the most highly protected tree in the rainforest, so frequently the surrounding trees are felled, leaving the Brazil nuts standing alone in an area used for agriculture. Although these trees can live for 200 years, the loss of the orchids and orchid bees mean the trees will not set seed and regenerate.

Is our planet big enough?

If moving to plant products sounds like a neat solution to our problems, consider the following difficulties.

We already cultivate a large proportion of the planet's surface for food, so there is not much room left for growing biofuels and other plant materials. Plants for biofuel are replacing rainforest (Figure 4.66) and also food crops.

Many of the remaining uncultivated areas are unsuitable for agriculture and/or are essential for wildlife conservation or water conservation.

Intensive agriculture is itself very energy-demanding (think of ploughing, harvesting and any fertilising of the soil).

The human population is increasing rapidly.

Countries such as India and China are rapidly catching up with Western levels of consumption.

Some of the plant alternatives do not have the same desirable qualities as the oil-based products.

Some plant alternatives may require the consumption of as much or more energy in production and transport of the product, compared with the oil-based products.

Figure 4.66 Palm oil plantations are replacing tropical rainforests on a large scale with the loss of all the forest biodiversity. As demand in the West for biofuels and other products has increased, the area used for palm oil plantations has risen drastically. In South East Asia, this area increased from 252 000 hectares in 1990, to 929 000 hectares in 2000 and to 2.15 million hectares in 2010. It is estimated that this will double again by 2020. Palm oil is now found in one in ten supermarket products.

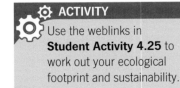

ACTIVITY
Use the weblinks in **Student Activity 4.25** to work out your ecological footprint and sustainability.

The richest one-sixth of the human population consumes approximately 80% of the world's resources. What is your own ecological footprint like? Your ecological footprint is the area of land you would require to support your current lifestyle. It assumes that the land you need is of average productivity for the Earth as a whole. It can then tell you how many planets like the Earth would be needed if everyone had an ecological footprint like yours. We need to consider how we can make changes to make our lifestyles sustainable.

4.5 On the brink

The primary threat to most species and habitats is human activity. Threats include land development (which causes habitat destruction, fragmentation and degradation), over-exploitation, introduction of alien species and pollution. Over the last 500 years, the extinctions of 869 species have been recorded. This rate (which is probably an underestimate) is much faster than expected from geological records.

The growing number of threatened species will not be saved from extinction merely by putting a fence around their remaining habitat and hoping for the best. As the golden lion tamarin story (Figure 4.67) illustrates, conservation of plants, animals or any other organisms requires a co-ordinated approach. Ideally, conservation management should be applied on site (*in situ*), protecting ecosystems and maintaining fragile habitats. In addition, vulnerable populations in the wild can be supported through the work of zoos and seedbanks. This is known as off-site (*ex situ*) conservation.

The role of zoos

The history of present-day zoos began in the 1750s, with the foundation of the first zoo in Europe by Emperor Franz Stephan in the grounds of Schönbrunn Palace just outside Vienna. London Zoo was founded in 1826, some six years before Charles Darwin set sail on the *Beagle*. Initially, the collection was privately studied by eminent scientists of the day, but opened its doors to the public in 1847. In those days before cameras and television, the public flocked to zoos, circuses and side shows to witness the wonders of the natural world. But almost nothing was known of these exotic animals' biology or behaviour and little heed was paid to their needs in captivity. They were usually kept alone in bare cages designed for maximum visibility by the public (Figure 4.68).

Today, zoos attract hundreds of millions of visitors each year throughout the world. They manage over one million vertebrates in over 1000 registered collections and most would claim to have a significant impact in conservation, research and education.

Figure 4.67 The golden lion tamarin (*Leontopithecus rosalia*) was on the edge of extinction in the early 1970s with only about 200 animals left in the wild. In 1974, the Golden Lion Tamarin Conservation Programme was set up. Scientists, conservationists and educators have worked together to protect and study the tamarin and its habitat so as to overcome the continuing threats to the species. Habitat management, captive-breeding, reintroductions to the wild, education and research are all part of this international rescue effort to save the species. The golden lion tamarin is a conservation success story. Assessed as Critically Endangered in 2000, it has now been downgraded to Endangered as a result of over 30 years of conservation efforts. There are now about 1600 golden lion tamarins in the wild, with about a further 475 in zoos involved in the conservation programme across the world.

> **WEBLINK**
>
> The International Union for the Conservation of Nature (IUCN) identified over 19 000 species known to be threatened with extinction in 2012. You can find out the current figure and exactly which species are threatened by visiting the IUCN Red List of Threatened Species website.

Figure 4.68 London's Royal Menagerie, 1812.

Centres for scientific research

One argument put forward for zoos is that they can play a vital role as research centres, enabling us to understand how to conserve particular species. The example below of mountain chickens at Jersey Zoo illustrates this.

How do mountain chickens breed?

The mountain chicken (*Leptodactylus fallax*) is not, as its name suggests, a bird, but is in fact one of the largest frogs in the world (Figure 4.69). It is found only on the Caribbean islands of Dominica and Montserrat. The mountain chicken's numbers have been declining because it has long been a national dish of the islands. It is hunted in large numbers for its meaty legs, which are used in traditional West Indian recipes. As their name suggests, their taste is somewhat like that of chicken.

Since 1995, over 75% of Montserrat has been engulfed by the continuing activity of the Soufrière Hills volcano. This volcano has destroyed much of the frog's remaining rainforest habitat and left the already threatened species in a very precarious situation.

Little was known about the reproductive biology of *L. fallax*. The main reason for this was that it leads a secretive, nocturnal life and so had proved difficult to study in the wild. In the battle to save the frog from extinction, more information was needed about its needs. In 1999, 13 frogs were taken to Jersey Zoo, allowing keeper scientists to find out more about their elusive lifestyle. A captive-breeding programme was set up: the zoo now has a captive-bred population of over 60 frogs, which also provides the potential for reintroduction in the event of extinction in the wild.

Q 4.39 **(a)** Suggest why the mountain chicken's underground nest-building behaviour (Figure 4.69) has evolved.

(b) What would be some of the complications of conducting a study on these frogs in the wild?

Figure 4.69 A *Leptodactylus fallax* – the chicken that's a frog. These frogs grow to 21 cm in length. **B** Spying on the foam nest. Unlike other frogs, mountain chickens lay their eggs in underground nests, covering them in foam.

Captive-breeding programmes

In the past, animal collections were unplanned and opportunistic. Zoos simply bought their animals from explorers and traders, who would return from expeditions in ships laden with crates of anything they could catch and manage to bring back alive.

Today, zoos seek to successfully breed the animals in their care through managed captive-breeding programmes. The aims of these programmes include:

● increasing the number of individuals of the species if numbers are very low

● maintaining genetic diversity within the captive population

● reintroducing animals into the wild if possible.

There are European Endangered Species Programmes for over 350 species, with about 350 institutions in Europe participating. Each species has an appointed co-ordinator/studbook-holder who advises on which animals should or should not breed and on the movement of animals between partner zoos, in order to maximise genetic diversity within the captive populations.

How genetic variation is lost

Genetic drift

In a small population, some of the alleles may not get passed on to offspring purely by chance, as shown in Figure 4.70. This change in the allele frequencies over time is known as genetic drift, and leads to a reduction in genetic variation.

WEBLINK
To find out more about captive-breeding programmes visit the European Association of Zoos and Aquaria website.

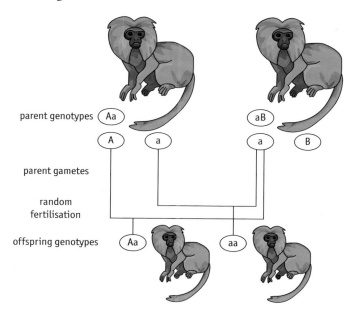

Figure 4.70 Genetic drift leads to less variation. In the example shown, the parents carry three alleles for a particular characteristic. By chance, one of the alleles is not passed on.

KEY BIOLOGICAL PRINCIPLE: VARIATION – THE KEY TO SURVIVAL IN A CHANGING ENVIRONMENT

Genetic uniformity, where individuals within a population have similar genotypes, can be an advantage in a stable environment. However, if the environment changes, a new disease emerges or a population moves, a genetically diverse population will be at an advantage.

Scientists interested in the preservation of species hope that there is sufficient genetic diversity in the population to ensure that some of the individuals can cope with any new conditions, allowing the population to survive. This is natural selection and it results in adaptation, the accumulation of genotypes favoured by the environment. No natural environment remains unchanged forever. It is therefore essential for long-term survival that populations should be able to evolve as a result of natural selection. Such evolution is unlikely unless there is genetic variation to enable evolution to happen, or the mutation rate is very high.

Inbreeding depression

In a small population, whether in the wild or in captivity, the likelihood of closely related individuals mating increases. This inbreeding causes the frequency of homozygous genotypes to rise, with the loss of heterozygotes. Inbreeding results in individuals inheriting recessive alleles from both parents and the accumulation of the homozygous recessive genotypes in the offspring. Many recessive alleles have harmful effects so **inbreeding depression** results. The offspring are less fit (less able to survive and reproduce). They may be smaller and not live as long and females may produce fewer eggs.

EXTENSION

Student Extension 4.4 allows you to discover some of the disadvantages and advantages of genetic uniformity in the 'ant supercolony story'.

Conserving genetic diversity

Conservation *in situ* (on site) to maintain the size of the wild populations is the best way to prevent genetic drift and inbreeding depression. However, *ex situ* (off-site) conservation can also play a role.

Keeping studbooks

The studbook for an individual species shows the history and location of all of the captive animals of that species in the places that are co-operating in an overall breeding plan. London Zoo, for example, keeps studbooks for several species, including the Sumatran tiger (Figure 4.71), the Fregate Island bettle and the Toco Toucan, and contributes to the studbook data for other species, such as the golden lion tamarin.

Studbooks provide the raw data upon which all the breeding plans are based – the scientists' understanding of genetics shapes the breeding plans themselves. The conservation scientists must ensure that genes from all the founder members of the population (the original group of individuals, usually wild-caught, on which the current population is based), or at least all remaining breeding adults, are retained and are *equally* represented in the subsequent generations (assuming that these founders are unrelated). This requires that individuals that breed poorly in captivity must be *encouraged* to breed, whilst those that are particularly good breeders must be *limited* in their breeding success.

This approach, using a breeding plan, is a very different principle from that followed by zoos of 50 years ago, when they simply raised their captive populations from the best breeders. This seemed like common sense, but it obviously reduced genetic variation and began a process that made it less likely that there could be a successful reintroduction to the wild.

Q 4.40 Suggest which of the named animals in Figure 4.72 is genetically most valuable for breeding purposes.

More and more, studbook records are being supported by techniques of cytogenetics – looking at the structure of chromosomes – and of molecular biology – studying the nature of the genes themselves. These additional techniques are important because studbook data may be incomplete

Figure 4.71 The Sumatran tiger is an endangered species that has been supported by a captive-breeding programme in zoos. Zoo managers make decisions about breeding these animals based on their studbook data.

WEBLINK

Visit the London Zoo (Zoological Society of London, ZSL) website to read more about their *in situ* and *ex situ* breeding programmes.

ACTIVITY

Investigate the animal dating agency for lemurs in **Student Activity 4.26**.

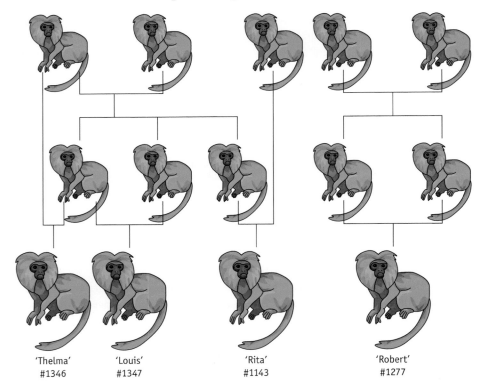

'Thelma' #1346 'Louis' #1347 'Rita' #1143 'Robert' #1277

Figure 4.72 Some of the breeding undertaken as part of the golden lion tamarin programme.

(since zoos in the past did not keep records as carefully as they do now). In addition, these new techniques can reveal if some individuals are more closely related than is desirable for breeding purposes.

Reintroducing animals into the wild

One role that zoos have in species conservation is captive-breeding for reintroduction. This involves breeding animals in captivity that are then returned to their native habitats. Clearly this can only work if the habitat is still intact so conservation of natural habitats is vital.

Reintroduction is a complicated process, particularly when species need to learn new skills before they have the ability to survive in the wild. However, there are cases in which zoos have successfully reintroduced captive-bred animals into the wild, as the example below illustrates.

Going, going ... saved

The Mauritius kestrel (*Falco punctatus*) was one of the rarest birds in the world, consisting of just four known individuals in 1974. The population was so small partly due to the extensive use of DDT as a pesticide to kill malaria-carrying mosquitoes. As the chemical worked its way up the food chain and accumulated, it was the kestrels as top predators that were most affected. Furthermore, the native habitat had become so degraded due to the removal of hardwood trees that it was not able to support a rapid increase in kestrel numbers without additional management techniques.

Carl Jones went to Mauritius soon after 1974 and realised that the species could be saved. With the backing of the Durrell Wildlife Conservation Trust, whose headquarters are at Jersey Zoo, he set up a captive-breeding centre on Mauritius where he took eggs from the wild birds and hatched them out in captivity (Figure 4.73). The young birds – fledglings – were then either returned to the wild as pairs of birds for rearing, or were 'hacked out'. 'Hacking out' involved them being taken out into the forest and gradually given more freedom and less food to encourage them to feed for themselves. Nest boxes were provided and natural cavities in trees modified to increase the number of suitable nest sites. In addition, predators, especially rats, were controlled at release and nest sites. As a result, there are now over 800 Mauritius kestrels slicing through the forest air.

Figure 4.73 Young captive-bred Mauritius kestrels, *Falco punctatus*.

 Q 4.41 What would be both the advantages and disadvantages of keeping the young Mauritius kestrels in open nest boxes in the forest?

 ACTIVITY
In **Student Activity 4.27** you analyse the reintroduction programmes for the Mauritius kestrel and ruffed lemurs.

CHECKPOINT
4.7 Summarise the roles of zoos in the conservation of endangered species.

Questioning the role of zoos

Many people are against keeping animals in zoos even though there is a growing awareness among those who run zoos of issues surrounding animal welfare. As we learn more about habitat protection and the need to conserve endangered species in their natural environments, the conservation role of zoos, and the justification for keeping wild animals in captivity, is being questioned (Figure 4.74).

The Born Free Foundation believes wildlife belongs in the wild. It is dedicated to the conservation of rare species in their natural habitats and the phasing out of traditional zoos. In 2000, it undertook a survey of the health status of UK zoos and published the report, Zoo Health Check 2000.

The Born Free Foundation report expresses concerns about the welfare of animals in zoos, having observed many animals exhibiting stereotypic behaviours, such as repetitive pacing up and down and chewing of bars. They found that up to one in five collections that qualify as zoos under the current legislation do not have a zoo licence. Perhaps more importantly from the animals' perspective, 95% of species in zoos are not endangered and are not part of European captive-breeding programmes. The Born Free Foundation would rather see animals being bred in protected habitats in the wild.

However, some endangered species only exist in zoos. They are extinct in the wild. Few people argue that they would be better off extinct than living in captivity.

UK zoos are regulated by the UK Zoo Licensing Act (1981), which aims to promote minimum standards of welfare, meaningful education, effective conservation, valuable research and essential public safety. The introduction of the European Zoos Directive (1999/22/EC) by the European Union places a greater emphasis on conservation, education, research and welfare.

Q 4.42 What arguments might be presented for and against keeping animals in captivity in zoos?

Figure 4.74 Keeping elephants in captivity – should zoos take into account that in the wild most elephants live in family groups and range over very large areas?

⚙ **EXTENSION**

Student Extension 4.5 allows you to consider some differing views on the role of zoos and debate the future of zoos.

The Millennium Seed Bank (MSB)

Plants are threatened worldwide by habitat destruction, climate change and over-harvesting. Protecting and managing habitats conserves plants *in situ*, but they remain at risk from man-made and natural disasters so there is also a role for *ex situ* conservation, using a variety of methods including seed banks and botanic gardens. The aim of the Millennium Seed Bank Project is to conserve seed samples from threatened species of plants, with 34 000 species banked in 2014. This represents 13% of the world's wild plant species. The aim of the project is to have 25% of species with bankable seeds by 2020. Seeds are collected around the world with a focus on plants and regions most vulnerable to human impact. Some are kept in their countries of origin and some are sent to the Millennium Seed Bank in the UK (Figure 4.75).

Most plants produce large numbers of seeds, so collecting small samples is unlikely to damage a wild population. Most seeds are small and easy to store, and can survive in a desiccated state for many years. But how do we know how long the seeds will live? What can we do to enable them to survive for a longer time?

The technology of seed preservation is improving all the time as more research is carried out. Seeds survive longer if kept dry and cool (Table 4.6). For every 1% reduction in seed moisture content, seed life span doubles, and for every 5 °C reduction in temperature, seed life span doubles. Once the seeds' identification has been verified, and they have been cleaned and dried, the seeds are stored at −20 °C.

Seed	Longevity (years)
Sunflower	165
Lettuce	447
Oilseed rape	843
Soybean	1122
Wheat	5079
Sugar beet	10 542

Table 4.6 Predictions of longevity under typical seed bank storage conditions.

Figure 4.75 The Millennium Seed Bank at Wakehurst Place in Sussex. The Millennium Seed Bank Project, a worldwide seed conservation network, aims to safeguard targeted wild plant species.

About a month after the seeds have been placed in cold storage, a sample is taken out and germinated on agar plates, to ensure that the seeds survive in the storage conditions. Germination is then tested about every 10 years to check that the seeds are still alive. If germination falls below 75%, then seeds will be grown to collect a new seed sample, which is then placed back into storage.

Some of the tests on seed viability have been carried out as school projects, supported by The Royal Botanic Gardens, Kew (Figure 4.76).

ACTIVITY
Use **Student Activity 4.28** to find out more about the Millennium Seed Bank project.

Figure 4.76 Schools have tested the effect of storage time on germination in a number of species as part of the Save Our Seeds project.

CHECKPOINT
4.8 So why are there so many different species? Use the relevant principles in this topic to answer this question.

ACTIVITY
Use **Student Activity 4.29** to check your notes at the end of this topic.

The seed collections are also used for research, habitat restoration and species reintroductions. Plants of *Silene tomentosa,* a species from Gibraltar thought to be extinct in the wild, were grown from seeds stored in the MSB and successfully reintroduced to Gibraltar. Seeds from UK native species, such as *Damasonium alisma* (starfruit), *Corrigiola litoralis* (strapwort) and *Apium repens* (creeping marshwort), have also been used for reintroductions.

TOPIC TEST
Now that you have finished Topic 4, complete the end-of-topic test. Congratulations – you have completed the AS.

REVIVING THE QUAGGA

It was thought that the quagga was a species similar to the plains zebra and it had become extinct in 1883. However the sources below suggest that this is not the case.

12:05 (GMT+2), *Tuesday, April 15, 2014*

QUAGGA REBREEDING: A SUCCESS STORY

By Keri Harvey

Until recently, it was believed that the last quagga died in Amsterdam Zoo in 1883. Today, however, this iconic animal is alive and back in the Western Cape. How was it possible to revive an animal from extinction? Keri Harvey speaks to the Quagga Project's Craig Lardner.

Contrary to popular belief, the quagga (*Equus quagga quagga*) is not a species in its own right. It is simply a subspecies of the Burchell's zebra or plains zebra … DNA analysis of quagga kept as museum specimens, has proven that the extinct quagga was in fact a Burchell's or plains zebra with a colour variation, in which some of its leg and rump stripes disappeared. This also means that Burchell's or plains zebra still carry genes from the extinct quagga …

Vanishing stripes

Why … the Burchell's or plains zebra lost some of its stripes is unclear, but … differing colouration seems to provide optimal camouflage: the quagga in each area blend better into their specific surroundings. Another purported reason for the quagga's vanishing stripes, apart from camouflage and hence protection from predators, is tsetse flies. It has been suggested that the zebra's stripes repel tsetse flies and so too the diseases they carry. Because the quagga lived outside the tsetse fly areas, the distinct stripes became obsolete.

… When it was discovered that the Burchell's or plains zebra is a DNA match for the extinct quagga, the project set about attempting to 'rebreed' the quagga … by selecting brownish zebra with reduced stripes and white tail bushes. In this way, the quagga genes could be concentrated to produce an animal that looks precisely like the 'extinct' quagga.

Only mitochondrial DNA was available from museum specimens and not nuclear and living DNA. For this reason, it was impossible to compare the rebred quagga to the original ones that became extinct. Nonetheless, the quagga in the Western Cape are believed to be the 'real thing', as it was in fact only coat pattern that distinguished a quagga from a Burchell's or plains zebra. Thus the Quagga Project seems to have succeeded in rectifying the tragedy that saw them being hunted to extinction

Extract of article published in the
South African Farmers Weekly *magazine and online*
http://www.farmersweekly.co.za

BIOLOGY LETTERS

A rapid loss of stripes: the evolutionary history of the extinct quagga

Abstract

Twenty years ago, the field of ancient DNA was launched with the publication of two short mitochondrial (mt) DNA sequences from a single quagga (*Equus quagga*) museum skin … (Higuchi et al. 1984 **Nature 312**, 282–284). This was the first extinct species from which genetic information was retrieved. The DNA sequences of the quagga showed that it was more closely related to zebras than to horses. However, quagga evolutionary history is far from clear. We have isolated DNA from eight quaggas and a plains zebra (subspecies or phenotype Equus burchelli burchelli). We show that the quagga displayed little genetic diversity and very recently diverged from the plains zebra …

… However, our results could be consistent with the quagga and the plains zebra being synonymized, as suggested earlier (e.g. Rau 1978; Groves & Bell 2004). Owing to priority, the correct name for plains zebras would thus be *E. quagga*, with, according to Groves & Bell (2004), five living and one extinct subspecies, the quagga (E. quagga quagga). …

… We estimate that this divergence took place in the Pleistocene, about 120 000 to 290 000 years ago… … (Dawson 1992). Therefore, the distinct coat colour of the quagga (Bennett 1980; figure 1) must have evolved quite rapidly. Existing plains zebras show a geographical gradient in coloration with progressive reduction in striping from north to south, which has been explained as an adaptation to open country and for which the quagga represented the extreme limit of the trend (Rau 1974, 1978). … Thus, the rapid evolution of coat colour in the quagga may be explained by either of two factors, or a combination of them: the disruption of gene flow owing to geographical isolation and/or an adaptive response to a drier habitat.

START BY REVIEWING THE SOURCE

First read the two extracts and think critically about them as sources of scientific information.

1. Criticise the reliability of each extract as a source of scientific information and justify your views with appropriate evidence.

2. Identify any statements in the *Farmer's Weekly* article that could be scientifically misleading.

REVIEWING THE BIOLOGY

1. Explain why people had considered the quagga to be a separate species.

2. Suggest what the term 'subspecies' means.

3. Drawing on your knowledge of adaptation, natural selection and evolution explain how the stripes will have disappeared from the populations of zebra known as quagga.

4. Explain with greater clarity the biology that underpins the sentence 'Because the quagga lived outside the tsetse fly areas, the distinct stripes became obsolete.'

5. Explain how DNA sequencing has altered the classification of the quaggas and plains zebra.

6. Suggest how the final sentence in the first paragraph of the *Farmer's Weekly* extract ('This also means that Burchell's or plains zebra still carry genes from the extinct quagga … ') contradicts what has been said earlier in the paragraph.

7. Suggest arguments that might be put forward to support the rebreeding of the extinct subspecies *E. quagga*.

8. Suggest how gene technology might be used to recreate the *E. quagga* subspecies.

The whole picture
As you read these articles think of everything you have learnt so far in the course and how it fits together to inform your understanding as a scientist. To help with your revision you could create a mind map for each topic and identify where ideas are revisited through the course. See the revision techniques in the exam support on the website.

Answers to in-text questions

Topic 1

Q 1.1 Movement of oxygen; carbon dioxide; and other products carried by blood relies on diffusion in animals with an open circulatory system; diffusion is only fast enough for small organisms;

Q 1.2 Blood can pass slowly through the region where gaseous exchange takes place; maximising the transfer of oxygen and carbon dioxide; and then be pumped vigorously around the rest of the body; enabling the organism to be very active;

Q 1.3 **(a)**

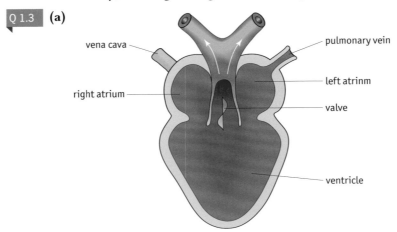

(b) Some mixing of oxygenated and deoxygenated blood; in the ventricle;

Q 1.4 **(a)** 3.5 stage micrometer units = 3500 µm
One eyepiece graticule unit = 3500/5
= 700 µm;

(b) (i) 140 µm
(ii) ×85;

Q 1.5 Thick layer of mainly elastic fibres; to allow expansion and recoil of artery; surrounded by thick layer of mainly collagen fibres; tough and durable to withstand high pressure;

Q 1.6 The blood pressure in the left ventricle falls below that in the aorta; leading to the closure of the semilunar valve between the left ventricle and the aorta; the blood pressure in the right ventricle falls below that in the pulmonary artery; leading to the closure of the semilunar valve between the right ventricle and the pulmonary artery;

Q 1.7 **(a) (i)** In atrial systole, the atrium contracts causing an increase in pressure in the atrium, forcing blood into the ventricular thus increasing the pressure in the ventricle; at the end of atrial systole the blood pressure starts to fall as the atrium relaxes;
(ii) In ventricular systole, the ventricles contract causing the pressure in the ventricles to rise, there is a pressure gradient between the ventricle and the atrium; as blood moves down this gradient it pushes against the atrioventricular valve causing it to close;
(iii) Blood moves from the high pressure in the ventricle to the low pressure in the aorta, opening the valve;
(iv) Blood entering the aorta from the ventricle causes the pressure in the aorta to rise;
(v) As the ventricles empty and relax the pressure in the ventricle falls below the pressure in the aorta, as blood moves down this pressure gradient it fills the flaps of the semilunar valve which closes;

(b) (i) open; **(ii)** closed;

(c) The shape of the graph would be the same as Figure 1.14 but with lower pressures, ranging from 0 to 30 mm Hg;

(d) One complete cardiac cycle lasts for 0.8 seconds; heart rate is calculated by dividing 60 seconds by the time taken for one cardiac cycle in seconds; 60/0.8 = 75 beats per minute;

Q 1.8 (a) The risk is 3811 in 727 724, or 3811 ÷ 727 724 = 0.005 or 0.5% chance of having a still birth in 2011;

(b) There were 6.6% fewer still births in 2012, 6.6% of 3811 is 251, so still births in 2012 was 3811 − 251 = 3560; the probability of having a still birth in 2012 is the number of still births divided by the total number of births 3560 ÷ 733232 = 0.005 or 0.5%;

Q 1.9

	Probability	
	Percentage	Per million people
Heart disease	0.23%	2300
Lung cancer	0.06%	600
Road accidents	0.006%	60
Accidental poisoning	0.0017%	17
Injury purposely inflicted by another	0.00049%	4.9
Railway accidents	0.00007%	0.7
Lightning	nearly 0	0.1

Q 1.10 (a) Ten students out of 1300 $\left(\dfrac{10}{1300}\right)$ is less than 1%; assumptions: the same exposure for each student; e.g. the same frequency of swimming for each student; the same length of each term with the same frequency of swimming each term; and the same viral load in and around the pool each term;

(b) Yes; risk of catching chlamydia in London in 2013 is 43 386 in 8 300 000

or 1 in $\dfrac{8\,300\,000}{43\,386}$

= 1 in 191 or 0.52%;

Risk of catching chlamydia elsewhere in England in 2013 is 208 755 in 53 500 000

or 1 in $\dfrac{53\,500\,000}{208\,755}$

= 1 in 256 or 0.39%;

Q 1.11 (a) positive correlation; no causal link; hot weather increases the number of ice cream sales and the number of people going swimming in shark-infested waters;

(b) positive correlation; no causal link; both these variables increase with age;

(c) positive correlation; causal link; smoking greatly increases the chance of developing lung cancer because tars and other substances in the smoke damage cells in the lungs;

(d) negative correlation; causal link; as the number of alcoholic drinks increases manual dexterity decreases as nerve messages are slower and less coordinated;

Q 1.12 No; the study only shows a correlation between smoking and lung cancer; evidence of how smoke damages living cells showed that smoking does cause cancer;

Q 1.13 It would not be valid to extrapolate these results to non-white populations because the cohort is not representative of the general population, which has a wide range of ethnic and racial groups; further studies have shown that the same results are true for most ethnic groups;

Q 1.14 It increases; greatly; for both females and males;

Q 1.15 Not necessarily; it might be that behaviours early in life greatly affect one's subsequent chances of developing cardiovascular disease;

Q 1.16 No; before the age of 75 males are more likely to die from cardiovascular disease than females are; male risk is 24 012 in 28 867 000 or 1 in 1202 or 0.08%, female risk is 10 124 in 29 038 000 or 1 in 2868 or 0.03%, after 75 years, the risks are similar for males and females, although men remain slightly higher risk; male risk is 1 in 46 or 2.17%, the female risk us 1 in 54 or 1.85%; the greater number of deaths among women is because they greatly outnumber men;

Q 1.17 The prevalence data in Table 1.1 is likely to be less reliable than the mortality data in Table 1.2; some people who have CHD may not recognise and report the disease;

Q 1.18 The data provide some support for the view that until the menopause a woman's reproductive hormones offer her protection from coronary heart disease in that deaths from cardiovascular diseases increase more steeply among women over the age of 50 than they do among men; but there are other possible explanations for this; so it would be premature to draw such a conclusion with any confidence;

Q 1.19 The decline may be due to fewer people having heart attacks and more people surviving if they had a heart attack;

Q 1.20 Blood pressure rises in the lungs due to a back-up of blood in the pulmonary capillaries; causing tissue fluid to build up within the lungs; impairing gas exchange;

Q 1.21 **(a)** 478 469 calories;

(b) 479 Calories;

Q 1.22 Carbon, hydrogen and oxygen are found in a 1 : 2 : 1 ratio;

Q 1.23 In galactose the —OH groups on carbon 1 and carbon 4 lie above the ring whereas in glucose the —OH groups on carbon 1 and carbon 4 lie below the ring;

Q 1.24

a sucrose

b maltose

c lactose

Q 1.25

Q 1.26 5;

Q 1.27 Energy requirements tend to increase up to the age of 18 years; after the age of 18 years energy requirements decrease; males' requirements remain static between 45 and 64 and then decrease with age; females' requirements decrease very little between 35 and 64; the decrease with age is partly due to a reduction in the basal metabolic rate (BMR) and to a reduced level of activity; on average, males have slightly higher requirements than females throughout life;

Q 1.28 For active 17 year old male: need to calculate BMR × 1.9; BMR = 12 900/1.4 = 9214;
9214 × 1.9 = 17 507 kJ/day;
For active 17 year old female: BMR = 10 300/1.4 = 7357;
7357 × 1.9 = 13 979 kJ/day;

Q 1.29 **(a)** BMI = 85/1.68² = 30.1, obese (although only just in obese category so may be considered as moderately obese);

(b)
$$30 = \frac{x}{1.91^2};$$
$$= \frac{x}{3.65};$$
$$30 \times 3.65 = x;$$
$$x = 109.4\,\text{kg};$$

Q 1.30 Waist-to-hip ratio 0.79; he is at low risk of heart disease;

Q 1.31 There is a positive correlation between increasing serum (blood) cholesterol and higher death rate for 30–39 year olds and 40–49 year olds; overall there is also an upward trend for 50–59 year olds, but the value for 6.21 is lower than for 5.69 serum cholesterol;

Q 1.32 The fact that pre-menopausal women generally have higher HDL : LDL ratios than men would be expected to lead to their having lower rates of coronary heart disease;

Q 1.33 LDL levels in the blood will fall; less saturated fat triglycerides in the blood to combine with cholesterol and protein to form LDLs;

Q 1.34 If the platelets do not stick to each other or damaged artery walls there is less chance of a clot forming that could block the artery;

Q 1.35 It is thought that the regular moderate consumption of alcohol in France may have a protective effect against CHD with the antioxidants in wine preventing the oxidation of LDL, which is involved in plaque formation; it may also be that the French diet contains a higher percentage of HDLs; there may be differences in other risk factors, for example obesity rates, physical activity;

Q 1.36 Nicotine patch user avoid the harmful effects of other chemicals in the smoke that could damage artery walls; may lead to person stopping smoking altogether.

Q 1.37 Six phenotypes; 3/3, 4/3, 3/2, 4/4, 4/2, 2/2;

Q 1.38 No increase or decrease in CVD risk compared to 'normal';

Q 1.39 Antioxidants protect against radical damage; radicals are highly reactive; and can damage many cell components; they have been implicated in the development of heart disease and some other diseases;

Q 1.40 People underestimate the risk associated with high cholesterol; they are unwilling and find it difficult to make the lifestyle changes needed to lower cholesterol;

Q 1.41 People who have had a heart attack may be more motivated to follow a strict diet; they probably started with higher blood cholesterol levels making it easier for them to reduce these; they may live in institutions where control over diet is greater;

Q 1.42 (a) To decide on the appropriate drug treatment her risk factors for heart disease should be considered; if she had high blood pressure, antihypertensive drugs might be prescribed to reduce her blood pressure; if she had high cholesterol, a drug to reduce blood cholesterol levels would be used; if there was a high risk of clotting then anticoagulant or platelet inhibitory drug treatment would be given; in deciding which particular drug to prescribe the benefits of the treatment would be considered along with any risk of side effects;

(b) No, it would not be sensible; for some people the risk of internal bleeding with aspirin may outweigh the benefit of reduced risk of cancer, particularly if their risk of bowel cancer was unknown or very low;

Q 1.43 Production of vitamin K is affected; vitamin K, calcium and thromboplastin are required to affect the conversion of prothrombin to thrombin, thrombin catalyses conversion of soluble fibrinogen into insoluble fibrin, this fibrous protein forms the mesh that traps blood cells in a blood clot;

Q 1.44 Cholesterol levels in the blood at the time of the stroke; blood pressure at the time of the stroke; report of the state of the artery when surgeons operated; genetic screening for the alleles that increase susceptibility to CVD;

Q 1.45 Eat a healthy diet; a diet low in saturated fats, low in salt, high in antioxidants; take regular exercise; do not smoke; avoid stress.

Topic 2

Q 2.1 Pathogenic microorganisms have time to multiply, resulting in illness or infection;

Q 2.2 Acid in the stomach kills the microorganisms;

Q 2.3 For A, SA = 6, Vol = 1, SA : Vol = 6; for B, SA = 24, Vol = 8, SA : Vol = 3; for C, SA = 96, Vol = 64, SA : Vol = 1.5;

Q 2.4 (a) Its surface area increases by a factor of 4;

(b) Its volume increases by a factor of 8;

(c) Its surface area to volume ratio halves;

Q 2.5 The surface area to volume ratio would continue to fall; the organism would not be able to exchange enough substances to survive;

Q 2.6 Hippopotamus;

Q 2.7 For D, SA = 34, Vol = 8, SA : Vol = 4.25; for E, SA = 28, Vol = 8, SA : Vol = 3.5;

Q 2.8 Volumes are all the same, 8; but the more elongated the block the greater the surface area and thus the larger the surface area to volume ratio;

Q 2.9 Tapeworm;

Q 2.10 Desiccation/dehydration problems; surface also has protective function;

Q 2.11 I;

Q 2.12 Kidney; intestines;

Q 2.13 They are carried in the bloodstream;

Q 2.14 Slower mixing at lower temperatures supports the model because movement of molecules in fluids is slower at lower temperatures;

Q 2.15 The kinks in the fatty acids prevent them lying very close together; this creates more space in which the molecules can move;

Q 2.16 (a) Diffusion;

(b) Active transport;

(c) Active transport;

(d) Facilitated diffusion/channel protein;

(e) Active transport;

(f) Osmosis;

Q 2.17 Two substrate molecules entering the active site in the first part of the figure, two substrates in the active site in the second part of the figure and one product molecule released in the third part of the figure;

Q 2.18 Decarboxylase – intracellular, catabolic; maltase – extracellular, catabolic; DNA polymerase – intracellular, anabolic; catalase – intracellular, catabolic; pancreatic lipase – extracellular, catabolic;

Q 2.19 A has a higher initial rate of reaction; $5/60 = 0.08 \, cm^3 \, s^{-1}$ (or $5 \, cm^3 \, minute^{-1}$); initial rate for B is approximately $1.8/30 = 0.06 \, cm^3 \, s^{-1}$;

Q 2.20 Salt is normally reabsorbed from sweat using the CFTR channel; with CF this does not function; so the salt is not absorbed making saltier sweat;

Q 2.21 It is a polymer; of nucleotides;

Q 2.22 A and G both have a two-ring structure whereas C and T have only one ring; the bases pair so that there are effectively three rings at each of the 'rungs' of the DNA molecule; making the molecule a uniform width along its whole length; the number of bonds that form between the molecules determines which bases pair; only two hydrogen bonds form between A and T, whereas three bonds form between C and G;

Q 2.23 (a) T A G G G A C T C C A G T C A;

(b) 46% of the DNA is C-G base pairs; therefore 23% is cytosine;

Q 2.24 Antisense strand: C T T C T A; mRNA: G A A G A U;

Q 2.25 (a) Messenger RNA;

(b) Transfer RNA;

Q 2.26 (a) TCA;

(b) AGU;

(c) UCA;

Q 2.27 A U G U A C C U A A G G C U A;

Q 2.28 5;

Q 2.29 UAC; AUG; GAU; UCC; GAU;

Q 2.30 Met, Tyr, Leu, Arg, Leu;

Q 2.31 The genetic code is degenerate, an amino acid may be coded for by several different triplets codes;

Q 2.32 Conservative replication: light and heavy bands at top and bottom of tube; fragmentary and semi-conservative replication: medium band in middle of tube;

Q 2.33

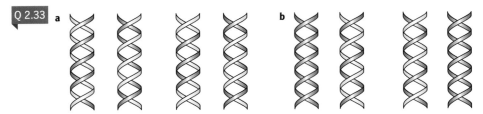

Q 2.34 If the fragmentary model was correct the DNA would all be medium density and appear as a band in the middle of the tube;

Q 2.35 G T C A G T C C G;

Q 2.36 **(a)** The sixth base is G rather than T;

(b) The second base is A instead of T;

(c) The fifth and sixth bases are inverted;

(d) Omission of the fourth base, A;

(e) An additional base, G, is added after the fifth base;

Q 2.37 In situations one and two both parents are probably carriers and there is a 1 in 4 chance that any child they have has CF; if the father in situation three is not a carrier then none of the woman's children will get the disease; they will all be carriers receiving a defective allele from her and a 'normal' allele from their father; if these children were to have children with a carrier there would be a 1 in 4 chance that each child would have CF;

Q 2.38 James and/or Margaret must be a carrier of the CF gene mutation; Polly has inherited the recessive allele from one of her parents; both Polly and Wilf are CF carriers and have passed the recessive alleles on to Frank;

Q 2.39 The test will not be completely reliable (it is only about 80–85% sensitive); there should not be false positives; but there will be false negatives where an individual has a CF mutation; but does not have one of the mutations that the test can detect;

Q 2.40 Yes; if the parents believe for religious or other reasons that abortion is wrong; if they consider that the risk of miscarriage is too high; if they want to have the baby even if it will have cystic fibrosis;

Q 2.41 Have a child accepting the 1 in 4 risk of it inheriting CF; use artificial insemination by donor to avoid the risk; use *in vitro* fertilisation and have the embryo screened before implantation; use prenatal screening and decide whether or not to continue with the pregnancy if the fetus has the disease.

Topic 3

Q 3.1 Cell surface membrane; smooth ER; mitochondrion; smooth ER; rough ER; nuclear membrane; nucleolus; nuclear membrane; rough ER; Golgi apparatus; cell surface membrane;

Q 3.2 Lysosome; free ribosomes;

Q 3.3 A – nucleolus; B – nuclear pore; C – nuclear membrane; D – mitochondrion; E – smooth endoplasmic reticulum;

Q 3.4 XYZ; XYZ; XYZ; XYZ; XYZ; XYZ; XYZ; XYZ;

Q 3.5 16 (= 24); 32 (= 25);

Q 3.6 A 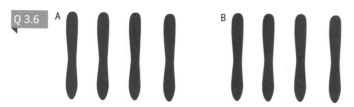 B

Q 3.7 **(a)** 75% tall purple and 25% short white;

(b) Long wings and broad abdomens, long wings and narrow abdomens, vestigial (short misshapen) wings and narrow abdomens, and vestigial wings and broad abdomens;

Q 3.8 Fruit flies have four pairs of chromosomes, the genes on each of the chromosomes are linked, with the same genes linked on each of the pairs;

Q 3.9 Colour blind girls can occur when a colour blind man has children with a heterozygous woman. The allele for colour blindness can then be passed to a daughter in the X chromosome from each parent;

Q 3.10 The family would not be free of haemophilia because their daughters would be carriers (heterozygous); half of the sons of the carrier daughters would be expected to have the condition;

Q 3.11 G1 = 8 hours; S = 4 hours; G2 = 8 hours; division = 4 hours;

Q 3.12 No distinct chromosomes are visible; protein synthesis is taking place;

Q 3.13 Ribosomes; endoplasmic reticulum;

Q 3.14 Condensing allows the DNA molecules to move around the cell without getting tangled up;

Q 3.15 The pores in the nuclear envelope are not large enough for whole molecules of DNA to pass through; once it disintegrates the chromosomes can move freely through the cell;

Q 3.16 Mitosis is a continuous process but a slide or photograph is a static image; the shorter the stage, the less likely a cell is to be in that stage when the action is 'stopped' as the slide is made or the photograph taken; it can be deduced that the fewer cells seen in any one stage, the quicker that stage is; anaphase is shorter than telophase because there is only one cell in anaphase and three in telophase; a larger sample would be needed to give a more accurate timing;

Q 3.17 **(a)** Asexual; sexual;

(b) Sexual; asexual;

(c) Sexual; asexual;

(d) Sexual; asexual;

(e) Asexual; sexual;

(f) Asexual; sexual;

(g) Asexual; sexual;

Q 3.18 Meiosis increases genetic variation increasing the chance of some cells having characteristics that will aid survival in the adverse conditions;

Q 3.19 G1; G2;

Q 3.20 The chemical messenger that carries information from the nucleus to the cytoplasm is mRNA; it codes for the proteins that determine the structure and function of the hat;

Q 3.21 The adult cell providing the genetic information to create Dolly was a specialised mammary gland cell; the successful birth of Dolly suggests that the cell must have contained all the information for making a complete organism;

Q 3.22 **(a)** Tissue; organ; organ; cell;

(b) Transport; photosynthesis; excretion; communication; or similar;

 Q 3.23 **(a)** Sepals and petals;

(b)

 Q 3.24

Bbb	BB bb bb (2)	Bb bb bb (1)
BbB	BB bb Bb (3)	Bb bb Bb (2)
Bbb	Bb bb bb (1)	bb bb bb (0)
bbB	Bb bb Bb (2)	bb bb Bb (1)

The number of alleles adding pigment is shown in brackets; the child with three of these alleles will have light brown eyes; there is a 1 in 8 chance of the couple having a child with brown eyes;

 Q 3.25 **(a)**

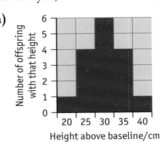

(b) Height shows continuous variation;

(c) **(i)** Parental phenotypes   Baseline height plus 30 cm
Possible gametes   **AB Ab aB ab**

(ii)

	AB	Ab	aB	ab
AB	AABB 40	AABb 35	AaBB 35	AaBb 30
Ab	AABb 35	AAbb 30	AaBb 30	Aabb 25
aB	AaBB 35	AaBb 30	aaBB 30	aaBb 25
ab	AaBb 30	Aabb 25	aaBb 25	aabb 20

A and B contribute 10 cm above baseline height while a and b add 5 cm to baseline height; the ratio of heights above baseline (cm) 20, 25, 30, 35, 40 in the offspring is 1 : 4 : 6 : 4 : 1;

 Q 3.26 1 and 2 are the result of genotype; 3 to 6 are environmental;

 Q 3.27 MSH activates melanocytes by attaching to MSH receptors on the cell surface; without these receptors, MSH has no effect and no melanin is made in the hair follicles;

 Q 3.28 In humans sunlight stimulates MSH receptor production; there is more sunlight in the summer;

Q 3.29 The tips are slightly cooler and so the enzyme remains active; melanin is made and tips are darker;

Q 3.30 Humans with this condition have heat-sensitive tyrosinase (like Himalayan rabbits and Siamese cats); so in the warmer parts of the body, such as the armpits, tyrosinase does not function and the hair is white; on the cooler surfaces of the body, tyrosinase works and the hair is dark;

Q 3.31 **(a)** The effects of diet may have caused epigenetic changes that were passed on via the nucleus of sex cells to subsequent generations; the chemical in the diet may have only occurred in sufficient concentrations when food is plentiful; causing changes to epigenetic markers on genes that are associated with control of blood sugar;

(b) X-linked epigenetic changes with inheritance via sperm;

Q 3.32 Each time DNA replicates before mitosis, a few errors will be made (mutations); the higher the rate of mitosis the more likely it is that mutations will occur in genes controlling cell activity, which can result in cancer;

Q 3.33 When the embryo grows into an adult, by mitosis, the cells giving rise to the ovaries or testes may have the DNA error; thus gametes with faulty DNA could form; the cancer-causing error could be passed on in these gametes to the next generation;

Q 3.34 **(a)** Less methylation of the oncogenes would mean increased transcription of the proteins that stimulate the transition between stages in the cell cycle, making the cycle more active with increased cell division;

(b) More methylation of the tumour suppressor genes would prevent transcription of the proteins that stop the cell cycle moving from one stage to the next, leading to increased cell division;

Q 3.35 If the single normal allele becomes damaged, a person with only one normal allele would no longer produce p53; whereas if someone with two normal alleles gets one of them damaged, they would still make p53 and be protected from cancer;

Q 3.36 If chemotherapy and radiotherapy work by activating the genes for p53; they will have no effect if the genes controlling p53 synthesis are not working.

Topic 4

Q 4.1 **(a)** A is a different species from B since 2% is a high percentage of genetic difference;

(b) A and C are either the same species or very closely related species as they can interbreed and produce fertile young;

(c) B and D are closely related, but distinct species as the males that result from interbreeding are infertile;

(d) E and F are the same species since there is no genetic difference between them;

Q 4.2 **(a)** Ensures that pollen comes only from a member of its own species;

(b) If the insect is not present (or becomes extinct) the plant will not be pollinated and could not reproduce;

Q 4.3 The woodpecker might be better adapted and displace the finch, or the woodpecker would find the niche already filled and not be able to survive;

Q 4.4 **(a)** Sight of the red spot on beak; chick obtains food from parent;

(b) Sight or sound of predator; draws predator away from young in the nest;

(c) Humidity of surroundings; woodlice will be less likely to dry out in damper conditions; in drier conditions they move around more so they are more likely to move to damper conditions; once in damp conditions they move around less so they stay in the more humid conditions;

Q 4.5 Active transport;

Q 4.6 Bumblebees with longer tongues tend to take nectar from flowers with longer corollas; this reduces competition between the different species; however, there is some niche overlap; bramble and white clover were used by a wide range of species;

Q 4.7 Brazil nut – hard coat to fruit; structural;
Agouti – sharp teeth; structural;
Euglossine bee – long specialised tongue; structural;
Orchid – bee attractant chemical produced; physiological;

Q 4.8 Ants gain protection; shelter;

Q 4.9 No; because there might be some disadvantages in having too long a tongue, for example it might be more likely to break or be difficult to carry in flight;

Q 4.10 1**A**; 2**B**; 3**C**; 4**D**; 5**E**; (3**B** and 2**C** are also acceptable);

Q 4.11 Frequency of homozygous recessive individuals, q^2 is 0.57, so the frequency of q, the recessive allele, $q = \sqrt{0.57} = 0.76$; $p + 0.76 = 1$, therefore $p = 1 - 0.76 = 0.24$; the frequency of the heterozygous individuals with genotype Ss is $2pq = 2 \times 0.24 \times 0.76 = 0.36$ or 36%;

Q 4.12 $q^2 = 1$ in 10 000 = 0.0001; so $q = 0.01$; and $p = 0.99$; $2pq = 0.0198$; 2% of the UK population will be carriers;

Q 4.13 (a) 81 of 243; 33% of student are homozygous recessive for the PTC allele; $q^2 = 81$ in 243 = 0.33; so $q = 0.57$; and $p = 0.43$; $2pq = 0.49$; half of these heterozygotes alleles will be recessive; 0.25; frequency of the recessive allele is 0.58, 58% of the alleles for this gene are recessive;

   (b) Yes; $q^2 = 65$ in 215 = 0.30; so $q = 0.55$; $p = 0.45$; $2pq = 0.49$; frequency of the recessive allele is 0.54 or 54%;

Q 4.14 A genetic mutation; this might work, for example, by making the mosquito's exoskeleton impermeable to the insecticide, or it might code for an enzyme that breaks down the insecticide; with a larger gene pool there are more likely to be individuals which are able to survive if conditions change; these survivors could give rise to a population which is adapted to the new conditions;

Q 4.15 (a) *Ranunculus*;

   (b) Meadow sweet and dropwort; they both belong to the same genus;

Q 4.16 *Chaetodon auriga*;

Q 4.17 They have a long extended part to their dorsal fins;

Q 4.18 Both woods have the same species richness, with eight species of tree; but the ancient wood has more evenness, i.e. the numbers of each species are more similar; the ancient wood is considered by ecologists to have greater biodiversity, despite (in our example) having the same number of tree species;

Q 4.19 Managed wood D $= \dfrac{9900}{3176} = 3.12$; ancient wood D $= \dfrac{9900}{1284} = 7.71$;

Q 4.20 Plants are easier to identify; a much greater proportion of the world's plants than animals have been identified; plant diversity is related to animal diversity;

Q 4.21 Hotspots appear to be in the warmer regions of the world; greater availability of sunlight and precipitation provide the best conditions for plant growth; plant and associated animal diversity will tend to be lower where conditions are less favourable; areas more recently glaciated will have lower diversity;

Q 4.22 (a) Nucleus;

(b) Chloroplasts;

(c) Golgi apparatus;

(d) Cell membrane;

(e) Vacuole;

Q 4.23 (a) One of the molecules will have to be inverted; so that the two –OH groups lie alongside each other and can react;

(b) It links the –OH group on the first carbon of one glucose with the –OH attached to the fourth carbon of another glucose;

Q 4.24 *Similarities*: both contain glucose, both contain 1,4 glycosidic bonds and hydrogen bonds; *Differences*: starch composed of a-glucose, cellulose b-glucose with each alternate glucose rotated through 180°; starch 1,4 and 1,6 glycosidic bonds, cellulose only 1,4 glycosidic bonds; starch amylose and amylopectin compact spiral shape held in place by hydrogen bonds, in cellulose, hydrogen bonds form between –OH groups in neighbouring cellulose chains, forming bundles called microfibrils.

Q 4.25 The cellulose glucose chain is similar to the chains running around the leg in nylon tights, the hydrogen bonds between the chains are weaker bonds and may have more 'give'.

Q 4.26

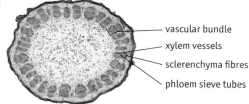

vascular bundle
xylem vessels
sclerenchyma fibres
phloem sieve tubes

Q 4.27 Release of enzymes from within the vacuole may digest cell contents;

Q 4.28 The specific heat capacity of a substance is the amount of energy in joules required to raise the temperature of $1\,cm^3$ $(1\,g)$ of water by $1\,°C$; it is very high for water because a large amount of energy is required to break the hydrogen bonds between water molecules;

Q 4.29 Active transport;

Q 4.30 Roots; fruits; seeds; tubers; corms; bulbs;

Q 4.31 To move the sucrose and other substances against a concentration gradient;

Q 4.32 *Similarities*: a liquid and all the particles it contains are transported in one direction through tubes; *Differences*: in plants there is no physical pump like the heart in the mammalian circulatory system, active transport of solutes into and out of the phloem sieve tubes creates a pressure difference that results in the mass flow along the tube;

Q 4.33 Strength suited to use; length; ease and reasonable cost of extraction; durability; resistance to decay;

Q 4.34 (a) (i) 2^{48};

(ii) 2, by the end of the 24 hours it will be 60% through the second division;

(b) 16×10^6;

Q 4.35 (a) (i) $2^{12} = 4096$;

(ii) $2^{14} = 16\,384$;

(b) (i) $\log_{10}$ for 4096 cells = 3.60;

(ii) $\log_{10}$ for 16 384 cells = 4.21;

Q 4.36 These parts come into greater contact with the soil; so are at greatest risk of microbial infection;

Q 4.37 He tested the drug on patients with symptoms of the disease; recorded any side effects; used standard procedure to discover the effective dosage; slowly increased the dose until patients experienced diarrhoea and vomiting and then reduced the dose slightly; recorded all results meticulously and published results;

Q 4.38 **(a)** It is difficult for those involved in the production of the new compound to be completely objective about it;

(b) Phase III is more rigorous; larger sample size; double-blind; use of statistics;

(c) Yes; because it is not known for certain if the new compound is effective; patients freely consent to participate in trials;

(d) The influence of mind over the body; the patient may have been getting better anyway due either to chance or their own immune response;

Q 4.39 **(a)** Scarcity of surface water; competition for these water bodies; high densities of aquatic predators; foam reduces risk of desiccation;

(b) Financial cost of researchers' travel, their food and accommodation; logistics of working in the field raise problems such as finding the nests in the first place; avoiding disturbance to egg-laying and care of young; the impact of the weather on conducting regular observations; liaison with local people;

Q 4.40 Robert; he is the only offspring from the two founders on the right; his parents are related so he is inbred, but he can be paired with an unrelated animal;

Q 4.41 *Advantages*: same environment as they will have to get used to after reintroduction; *Disadvantages*: risks of disease and death due to weather; parasites; predators;

Q 4.42 *Arguments for*: role in academic research; captive breeding with careful use of studbooks to maintain genetic diversity; reintroduction programmes; education; *Arguments against*: animals are exhibited on the basis of their 'crowd pulling' power, rather than on their endangered status; animals are kept in inappropriate conditions; both in terms of their physical environment (poor floors in cages, for example); and their mental and social well-being (limited feeding stimuli or atypical social grouping); capture of animals for exhibition may seriously deplete wild populations; reintroduction of species to the wild cannot be guaranteed.

Picture credits

The authors and publisher would like to thank the following individuals and organisations for permission to reproduce photographs:

(Key: b-bottom; c-centre; l-left; r-right; t-top)

123RF.com: Marty Wakat 165; **Alamy Images:** Arco Images GmbH 178, BSIP SA 97, Danita Delimont 198l, Edward Parker 150l, 175br, John Warburton-Lee Photography 157, Juniors Bildarchiv GmbH 156bl, Lumi Images 142, Nigel Cattlin 185, Royal Geographic Society 41, sciencephotos 29, Tim Gainey 154t, WilliamRobinson 164l; **Angela Hall:** 183; **Ardea:** Bill Coster 155, Valerie Taylor 166bl; **Corbis:** 2/Tom Brakefield/Ocean 200, Emrah Turudu/145/Ocean 143b, Historical Picture Archive 197b, Jerome Gorin/PhotoAlto 202l, Peter Johnson 170b; **Digital Vision:** 180t; **DK Images:** Nigel Hicks 150c, Steve Shot 144r; **Durrell Wildlife Conservation Trust/Gerardo Garcia:** J Copsey 198r; **FLPA Images of Nature:** Reinhard Dirscherl 166tc; **Fotolia.com:** Axel Gutjahr 191bl, Bombaert Patrick 194, caan2gobelow 166cl, creativenature.nl 104br, Dinadesign 189bl, martin1985 197t, Paul Hakimata 20, S.R. Miller 154b, visceralimage 153l, volff 34; **Getty Images:** 97 51, Bob Elsdale 106t, Yorick Jansens/Stringer 39; **Marine Themes:** Kelvin Aitken 168c; **Nature Picture Library:** Nick Garbutt 201; **Pearson Education Ltd:** 33l; **Photoshot Holdings Limited:** NHPA 150r, 164r; **Phototake, Inc:** C. James Webb 91; **Randy L. Jirtle:** Randy Jirtle 144l; **Rex Features:** 130, FLPA 156tl; **Royal Botanic Garden Kew:** Copyright Board of Trustees of the Royal Botanic Gardens, Kew 203t, 203b; **Science Photo Library Ltd:** A Stenning 181, 58r, 81r, 129, 191cr, Alex Bartel 176, Andy Walker, Midland Fertility Services 104tr, 126, 138l, Antonia Reeve 53, Astrid & Hanns-Frieder Michler 61, Biografx/Kenneth Eward 77, Biophoto Associates 10, 15b, 107, 177, 179, 186, Biophoto Associates 10, 15b, 107, 177, 179, 186, BSIP/Amelie-Benoist 101, BSIP, B. Boissonnet 27, BSIP, L. Souci 98, BSIP, Sercomi 106b, Carlos Munoz-Yague/Eurelios 189t, Claude Nuridsany & Marie Perennou 188t, CNRI 138cl, Cordelia Molloy 35, Don W. Fawcett 65, 68, Dr Gopal Murti 122cl, Dr Jeremy Burgess 118, 124c, 139, 170t, 180bl, 193, Dr Keith Wheeler 217, Dr Scharf 160, Dr Yorgos Nikas 111, Dr. Don Fawcett 109, Dr. Elena Kiseleva 88, Eye of Science 138c, 138b, Garry de Long 182b, Geoff Kidd 127tr, George Bernard 125r, Jame King-Holmes 128t, James Stevenson 147, Laguna Design 65cl, Manfred Kage 124b, Martin Oeggerli 128bl, Maximilian Stock Ltd 50, Pascal Goetgheluck 104tl, Power and Syred 138cr, 182c, Pr. G Gimenez-Martin 120, 122tr, 122br, 123t, 123bl, 123br, Rosenfeld Images Ltd 33r, 127b, Science Source 80, 81l, Sinclair Stammers 125c, Sovereign, ISM 17, Steve Gschmeissner 15c, 58l, 104bl, 138r, Susumu Nishinaga 16, 59, The Biocomposites Centre/Eurelios 188b, Visuals Unlimited, Inc./Carolina Biological Supply Co 14r, Visuals Unlimited, Inc./Frederick C. Skvara 14l, Volker Steger 189br; **Anne Scott:** 175t, Mark Tolley 4t, 4b, Peter Kempson 5t, 5b; **Shutterstock.com:** BMJ 143c, Chris Fourie 119, hohotun4ik 56, Hugh Lansdown 153r, Paul Banton 202r, Sarah-Jane Walsh 125l, stefanolunardi 2; **The College of Physicians of Philadelphia:** Evi Numen 143; **Veer/Corbis:** Samot 3, Stephan 153c, Wong Sze Fei 196, Yelena Kovalenko 47, Zafer KIZILKAYA 166tl

Cover images: *Front:* **Science Photo Library Ltd:** Frans Lanting, Mint Images

All other images © Pearson Education

We are grateful to the following for permission to reproduce copyright material:

Figures
Figure on page 13 adapted from *Anatomy & Physiology: The Unity of Form and Function*, McGraw-Hill Higher Education (Saladin,K); Figure on page 54 from Rare Variants in NR2F2 Cause Congenital Heart Defects in Humans Al Turki, Saeed et al. The American Journal of Human Genetics Volume 94 Issue 4 574–585, Granted under Creative Commons Licence Attribution 3.0 Unported (CC BY 3.0) Screenshot on page 102 from an on-line resource on breathing and asthma produced by the Association of the British Pharmaceutical Industry (ABPI) to support biology education in UK schools reproduced with permission.

Text
Quote on page 54 from Saeed Al Turki, first author from the Wellcome Trust Sanger Institute Press release published on the Wellcome Trust Sanger Institute website at http://www.sanger. ac.uk/about/press/2014/140408.html, Wellcome Trust Sanger Institute with permission; Quote on page 54 from Dr Matthew Hurles, senior author from the Wellcome Trust Sanger Institute. Press release published on the Wellcome Trust Sanger Institute website at http://www.sanger. ac.uk/about/press/2014/140408.html, Wellcome Trust Sanger Institute with permission; Article on page 103 adapted from Childhood obesity is fuelling asthma epidemic, *Daily Express* 02/07/2014 (Willey, J); Quote on page 130 from *Stem cell research: Medical progress with responsibility* a report from the Chief Medical Officer's expert group reviewing the potential of developments in stem cell research and cell nuclear replacement to benefit human health, Crown Copyright: Contains public sector information licensed under the Open Government Licence v3.0; Quote on page 145 from Dr Branwen Hennig Maternal nutrition at conception modulates DNA methylation of human metastable epialleles by Paula Dominguez-Salas, Sophie E. Moore, Maria S. Baker, Andrew W. Bergen, Sharon E. Cox, Roger A. Dyer *Nature Communications* published by Nature Publishing Group Apr 29, 2014 Copyright © 2014, Rights Managed by Nature Publishing Group; Extract on page 148 from Effect of Trehalose on protein structure, *Protein Science*, 01/01/2009, 18(1): 24–36 (Nishant Jumar Jain and Ipsita Roy), Protein science: a publication of the Protein Society by Protein Society; reproduced with permission of Cold Spring Harbor Laboratory Press in the format republish in a book via Copyright Clearance Center; Article on page 204 from Quagga rebreeding: a success story, *South African Farmers Weekly Magazine* 15/04/2014 (Keri Harvey), http://www.farmersweekly.co.za, Farmers Weekly with permission; Article on page 205 from A rapid loss of stripes: the evolutionary history of the extinct quagga, http://rsbl.royalsocietypublishing. org/content/1/3/291 DOI: 10.1098/rsbl.2005.0323, published 22 September 2005, Jennifer A. Leonard, Nadin Rohland, Scott Glaberman, Robert C. Fleischer, Adalgisa Caccone, Michael Hofreiter, copyright © 2014, The Royal Society.

Every effort has been made to contact copyright holders of material reproduced in this book. Any omissions will be rectified in subsequent printings if notice is given to the publishers.